T0245958

Christine Figgener

Translated by Jane Billinghurst

MY LIFE WITH Sea Turtles

A MARINE BIOLOGIST'S QUEST TO PROTECT ONE OF THE MOST ANCIENT ANIMALS ON EARTH

DAVID SUZUKI INSTITUTE

GREYSTONE BOOKS

Vancouver/Berkeley/London

First published in English by Greystone Books in 2024
Originally published in German as *Meine Reise mit den Meeresschildkröten*,

24 25 26 27 28 5 4 3 2 1

Greystone Books Ltd.
greystonebooks.com

David Suzuki Institute
davidsuzukiinstitute.org

Cataloguing data available from Library and Archives Canada
ISBN 978-1-77840-058-2 (cloth)
ISBN 978-1-77840-059-9 (epub)

Proofreading by Jennifer Stewart
Indexing by Cameron Duder
Jacket and text design by Fiona Siu
Jacket photograph by Melinda Moore/Getty Images

All photographs by Christine Figgener or from her personal collection. Specific photo credits for 1 top and 7 bottom, Andrey Castillo MacCarthy; 2 top, Cody Glasbrenner; 3 bottom, Lisette Heijke; 4 top, Brianna Myer; 5 top left, Olga ga/Unsplash; 5 top right, Tom Doyle; 6 top and bottom, Monroe County Public Library, Florida Keys History Center, Online Digital Collection

Printed and bound in Canada on FSC® certified paper at Friesens. The FSC® label means that materials used for the product have been responsibly sourced.

Greystone Books thanks the Canada Council for the Arts, the British Columbia Arts Council, the Province of British Columbia through the Book Publishing Tax Credit, and the Government of Canada for supporting our publishing activities.

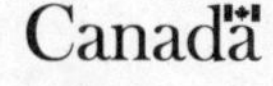

Canada Council for the Arts Conseil des arts du Canada

Greystone Books gratefully acknowledges the xʷməθkʷəy̓əm (Musqueam), Sḵwx̱wú7mesh (Squamish), and səlilwətaɬ (Tsleil-Waututh) peoples on whose land our Vancouver head office is located.

For Jairo, who loved sea turtles just as much as I do, and for my niece Charlotte, who stands right at the very beginning of her life journey

CONTENTS

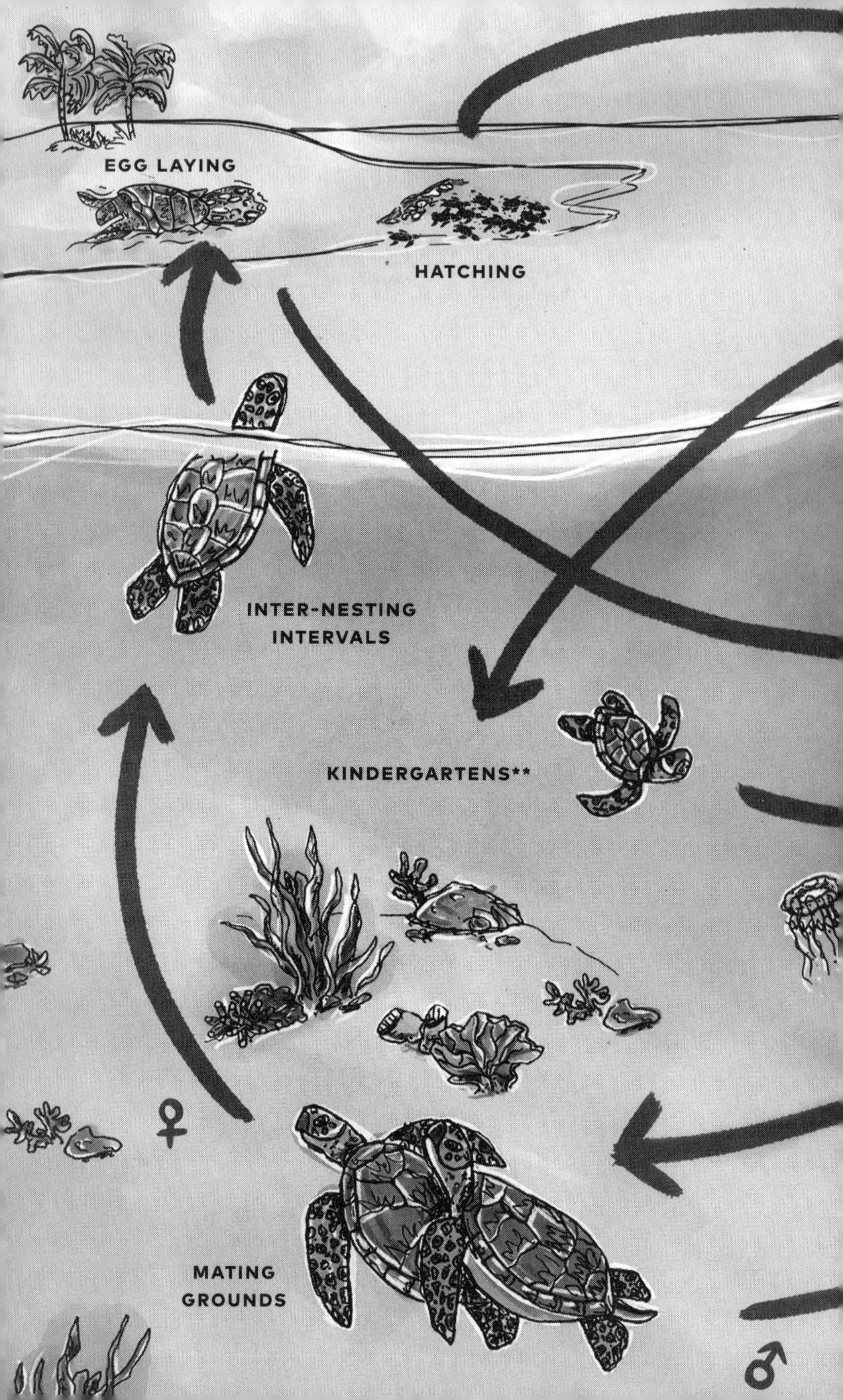
EGG LAYING
HATCHING
INTER-NESTING
INTERVALS
KINDERGARTENS**
♀
MATING
GROUNDS
♂

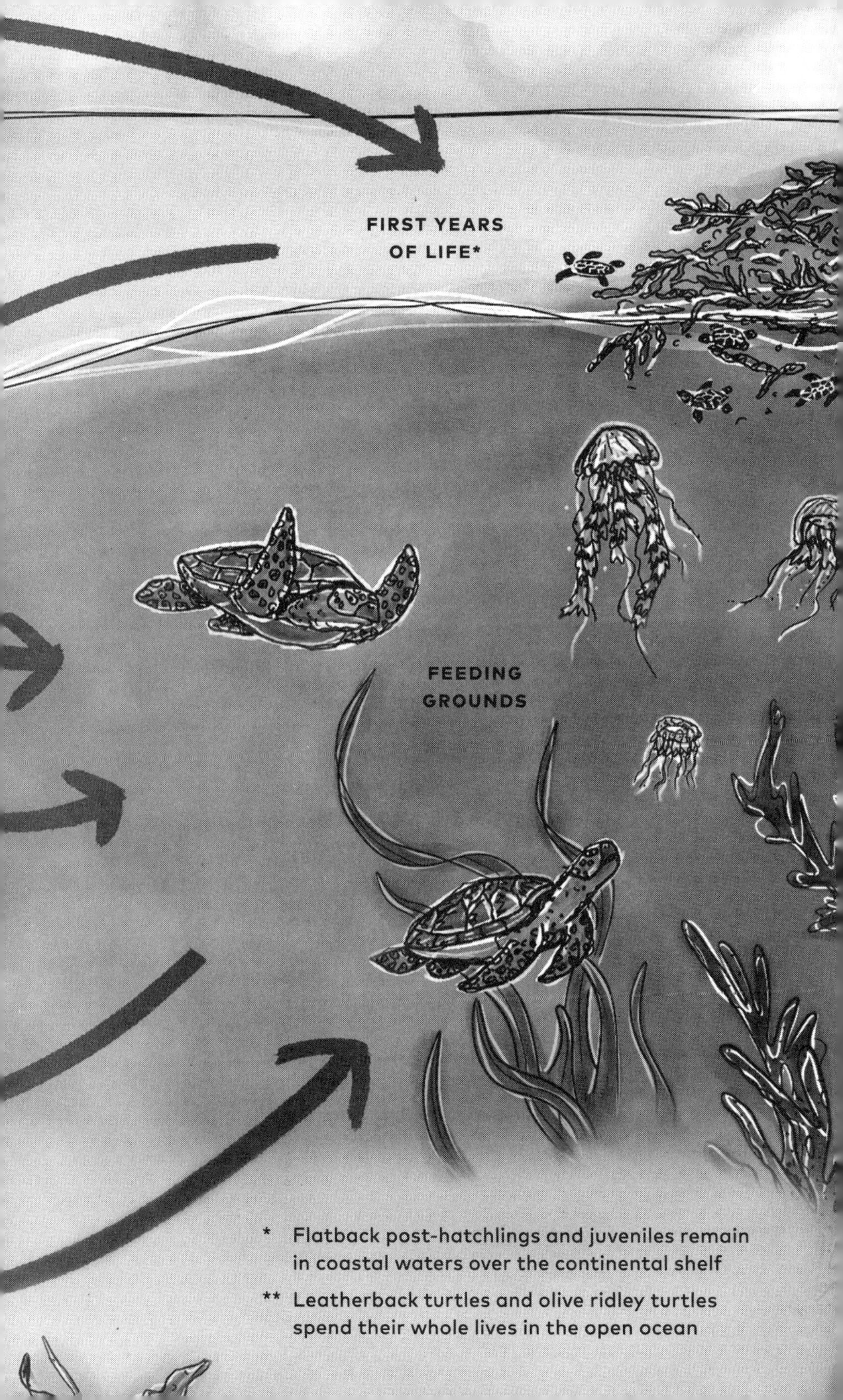
FIRST YEARS
OF LIFE*
FEEDING
GROUNDS
* Flatback post-hatchlings and juveniles remain in coastal waters over the continental shelf
** Leatherback turtles and olive ridley turtles spend their whole lives in the open ocean

ANATOMY OF A SEA TURTLE

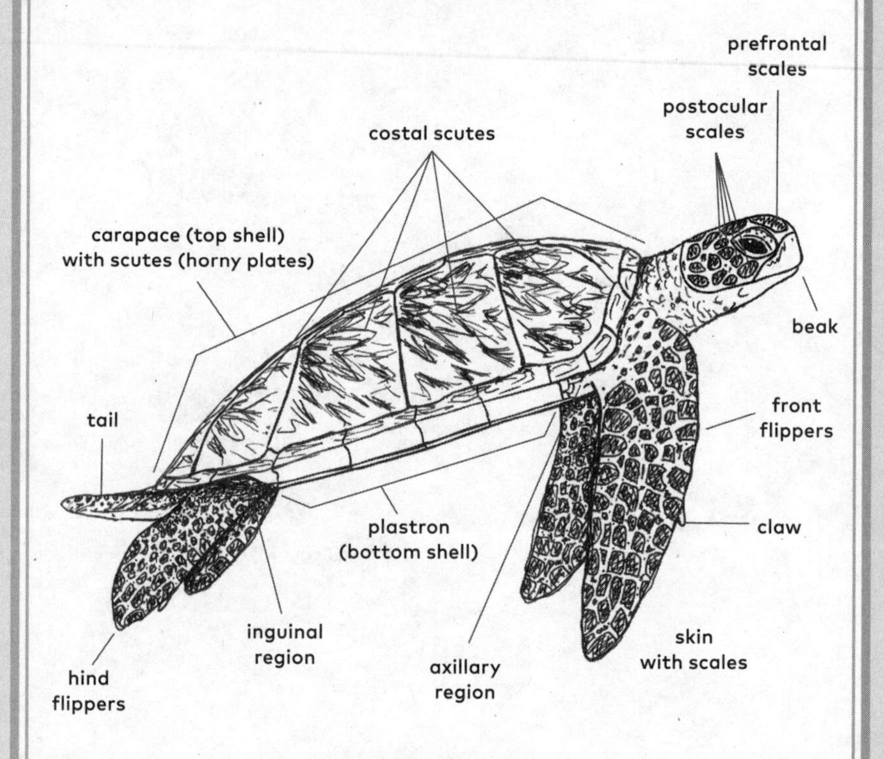

PROLOGUE

I'VE BEEN WALKING in the dark along this deserted beach for several hours. An incredible starry sky stretches out above me. The air feels heavy in my lungs; it smells of salt and musty wood with sweet, floral overtones. An impenetrable black wall of jungle looms up on one side of me, while on the other, I have to steer clear of waves from the Caribbean Sea. It's still warm even though it's after midnight, and under a backpack full of research equipment, my dark, long-sleeved sweatshirt is sticking to my back.

Even though my eyes have adjusted to the darkness, I'm stumbling along rather than walking. The beach is a cluttered obstacle course: downed trees, small branches, and dunes and dips everywhere in the soft, damp sand. I am constantly scanning the water's edge. Every once in a while, I spot an elongated shadow, and a shot of adrenaline courses through me. So far, every one of these shadows has turned out to be a tree trunk or stump, and I wonder if I'm going to recognize what I'm looking for if and when the time comes...

Behind me, my even less experienced companion, Michael, is sweating and tripping as he tries to keep up with me. After almost falling for the umpteenth time, he swears violently. "Why do we have to walk around in the dark? Why can't we just use our flashlights?"

I half turn toward him. "Sea turtles like darkness when they lay their eggs. When they climb up the beach, they orient themselves by the contrast between the brightest parts of the sky—usually the stars and the moon—reflecting off the water and the pitch black of the beach vegetation," I explain. "If we used white light, we could scare them away and they wouldn't lay their eggs." That seems to satisfy him, and I direct my attention toward the ocean once again.

For the past two weeks, I've been here in Costa Rica taking part in a project to protect leatherback turtles, and even though I pretend to Michael that I know what I'm talking about, I've yet to see a single one. I've been at the research station learning everything I can about the biology, ecology, and conservation of sea turtles—including how to collect data and what we need to do to protect nesting mothers and their eggs from poachers. So far my training has been purely theoretical as we've not come across any nesting females during our beach patrols. My training period is over. I've been thrown into the deep end and now here I am stumbling along the beach with a single volunteer to "save a few sea turtles," as the biologist in charge cheerfully put it to motivate us.

Sweat is running slowly from my butt down my legs, and I can feel the sandflies that have found their way under my long pants and are now helping themselves to a meal of blood. Perhaps next time I should pay attention to the recommendation to wear long socks. My thoughts wander to my bed and the safety of my mosquito net back at the research station. How much longer is my shift going to last? I peek at my watch. The display, fogged up from the inside because of the humidity, tells me that I still have two hours to go. It will be four in the morning before I'm finally able to creep, dead tired, into bed.

Suddenly my subconscious pulls my thoughts back to the beach. Did that dark log out in the waves just move? I stop abruptly and hold my breath. My companion crashes into me. "What's going on?" he wants to know.

"I think that's a turtle over there," I whisper. The log does indeed seem to have pushed itself somewhat farther up the beach. I take a couple of steps in its direction, and now I can see it clearly: the silhouette of an enormous sea turtle beginning to emerge from the shallow water. Waves are washing over her, and the silvery light of the moon is reflecting off her smooth shell. She keeps lifting her head as though trying to get her bearings.

My heart begins to beat wildly and my mind races. Frantically, I try to recall everything I learned during training. As information swirls in my brain, one detail in particular rises to the surface: a sea turtle is easily spooked, especially in the early stages before laying her eggs, and will crawl back into the water if she feels uneasy. The first thing we should do, therefore, is retreat. I tug on Michael's sleeve, and we walk a short distance higher up the beach, where the trees are casting dark shadows. I tell him we'll need to wait until the female begins to dig her nest. Only then will it be safe for us to approach.

Over the roar of the surf I hear the turtle's long flippers slapping the sand, and her grunts and groans as she pulls herself laboriously up the beach. At some point, the noise changes and it sounds as though sand is being thrown around. I creep forward under cover of darkness. The female appears to have found a suitable spot and is now busy preparing the site for nesting. For the next half hour, I leave the shadows at regular intervals and cautiously approach her from behind to see how she's doing. Time seems to stand still as I worry that the turtle might abort her nest building despite our precautions.

At least the long wait gives me time to get my nervousness under control and prepare our equipment. I take a data sheet out of the backpack and fill in the date and time. Then I scan the vegetation line with my red light until a white reflector gleams in the darkness. I walk up and read the number 25 written on a tree above it. Over the past couple of weeks, starting at the north end of the beach and moving south, we painted numbers on trees from 1 to 160 every 50 meters (55 yards) to help orient us as we gather our data. With the waves crashing behind us and marker 25 to our left, the turtle must be directly in front of the tree with the number 24 or maybe between 24 and 25. I make a careful note of that as well. We are going to need all this information later for our statistics. We also need it so we can find the nest again when we return to see how many of the eggs have hatched.

Finally, the huge body of the turtle stops moving, and she begins to dig her nest with her hind flippers. We both carefully creep forward to get closer to her. Now, for the first time, it's okay for us to shine our flashlights in her direction—but still using only red light and only on her back and tail. She is most sensitive to what's going on around her before she starts laying her eggs. As soon as she lays her first eggs, she will fall into a sort of nesting trance—and once in their trance, the females of most species are not easily disturbed.

I can't believe how big she is, I think, as I settle down on my stomach on the sand behind the turtle and run the beam of my flashlight up and over her back. From head to tail, she is longer than I am tall (I am 1.7 meters/5.6 feet!) and must weigh between 300 and 600 kilograms (660–1,320 pounds). As I will learn over the coming years, that's nothing out of the ordinary for a leatherback turtle, but my first sighting takes my breath away. I'm also fascinated by how gracefully she moves her hind flippers despite her bulk and her rather awkward appearance on land—she is using her flippers like hands to remove sand from her half-finished nest. Amazing!

Michael and I keep an awestruck silence as we watch the hind flippers at work. First one, then the other slides slowly into the ever-deepening egg chamber, the rear edge of each flipper scraping the sand as it reaches for the opposite side of the hole, and then the flipper bends to carry the sand up and out of the nest. The turtle then tosses the sand to one side with an expert flick of her flipper. It is a hypnotizing sight and might even have lulled me to sleep if I hadn't been so pumped full of adrenaline.

After watching for a few minutes, not moving a muscle, I notice something metallic flashing every time the female switches flippers. When I look more closely, I see a metal tag attached to the base of each flipper on the side closest to the tail. I'm briefly overtaken by a feeling of relief. This turtle has already been identified and tagged. Today is not the day I am going to have to put my sparse and, up until this point, purely theoretical knowledge of tagging sea turtles into practice. I pull on a pair of latex gloves and hand the clipboard with the waterproof data sheet to Michael.

Without touching either the turtle or her tail, I carefully rub the sand off both metal tags with one finger and read out the numbers and letters: "V1858, V1357." I get Michael to repeat the serial numbers back to me. They are correct. We have verified two more important pieces of information. So far all is going well, and that helps me calm down.

I now turn my attention to the surroundings because I need to decide whether the nest can be left alone or whether the eggs should be moved. If the latter, that means I will have to either hide the eggs in a more secure spot on the beach or move them to our hatchery a few miles away. We are about 30 meters (100 feet) above the last high tide line and about 3 meters (10 feet) from the trees bordering the beach. The turtle did not hit any tree roots or water as she was digging. Also, there are no homes along this section of the beach. All the signs bode well, and for the time being, the eggs appear to be safe here in the nest the turtle has dug.

I grab the clicker counter from the backpack and let Michael know what's going to happen next. "You'll know when she's finished digging when she scrapes a bit more at the side of the nest with one flipper but doesn't remove any sand. She'll then pull her flipper out of the nest and use it to cover her tail. That's your cue she's going to start laying her eggs." Michael will have to pay close attention, because his task is to count every egg that drops out of her cloaca, the opening at the base of her tail.

We don't have to wait long. After a few more minutes, the female pulls her flipper out of the nest and places it over her tail. Gently, Michael pushes his hand under the flipper and lifts it so he has a clear view into the egg chamber. I can see the female's flipper pressing down as she starts to push with her whole body, and then two white eggs about the size of billiard balls fall into the nest.

While Michael counts carefully, I pull out the most expensive piece of equipment in my backpack. It's a scanner I will use to search the turtle for embedded microchips. To my delight, I get a beep right away when I slide the scanner over her shoulder, and a number appears on the display panel. More information for our data sheet!

I look over Michael's shoulder. "Has she started to lay her false eggs yet?" He says she has and now I, too, see the smaller eggs lying in the nest. They vary in size from tennis balls to peas and everything in between. These false eggs or SAGS (short for shelled albumen globes) are simply egg white enclosed in a shell. They presumably serve as placeholders so that when the baby turtles hatch later on, they can interact with each other and coordinate their movements more easily. The false eggs create extra pockets of space by slowly deflating while the true eggs are incubating. We watch together as the last eggs plop into the nest. When the female is finished, she moves the flipper she had positioned over her tail into the nest and presses down on the eggs. Then she pulls that flipper out, places it next to the nest, and slides her other hind flipper inside the nest along with a little sand. She presses down once more on the eggs and compacts the sand. If I still needed to attach tags to her flippers, now would be the perfect time. She repeats the process until the eggs are completely covered and the nest chamber is full of sand.

All we have to do now is measure our leading lady. Generally we use a tape measure placed over the curve of the turtle's back to measure the curved length and width of the top shell, or carapace, excluding the head and tail. With a leatherback turtle, this takes two people as most leatherbacks are longer and wider, even, than the span of most people's arms. Apart from that, it's quite uncomfortable to be practically doing the splits over a turtle's back. That's something I will learn the hard way with other turtles later on. Tonight, however, we have no trouble taking our measurements even though the female has begun to camouflage her nest and we have to be careful that her enormous front flippers don't knock us off our feet. With a carapace length of 1.62 meters (5.31 feet), she is definitely an impressive specimen; the average for a Caribbean leatherback turtle is 1.55 meters (5.09 feet). After we have made a note of her measurements, along with the number of eggs (ninety-six!), our initial work is done. We gather up our equipment and sit down on the sand a short distance from the nesting area to watch as the turtle makes final sweeps with her flippers. I notice how I am gradually relaxing, and how, finally, adrenalin is slowly giving way to endorphins. My first leatherback turtle! Absolutely breathtaking!

While the female grunts as she's throwing sand around to camouflage the nest and cover her tracks, I allow my thoughts to wander. Never in my wildest dreams did I imagine that I would one day move from Germany to Central America to research and protect sea turtles. But now here I am, sitting on the other side of the world under a starry sky so bright that I've seen its equal only a handful of times in my life. In the darkness, I can still make out the silhouette of the huge leatherback that has just laid her eggs in my presence. I pinch myself. But this is no dream. I can feel the breeze from the ocean on my skin and am aware of the smell of algae and cloacal fluid wafting toward me from the mother turtle.

Looking back as I write this, I can truly say that my encounter with this single leatherback turtle kick-started the greatest adventure of my life.

OLIVE RIDLEY TURTLE

VU

Lepidochelys olivacea

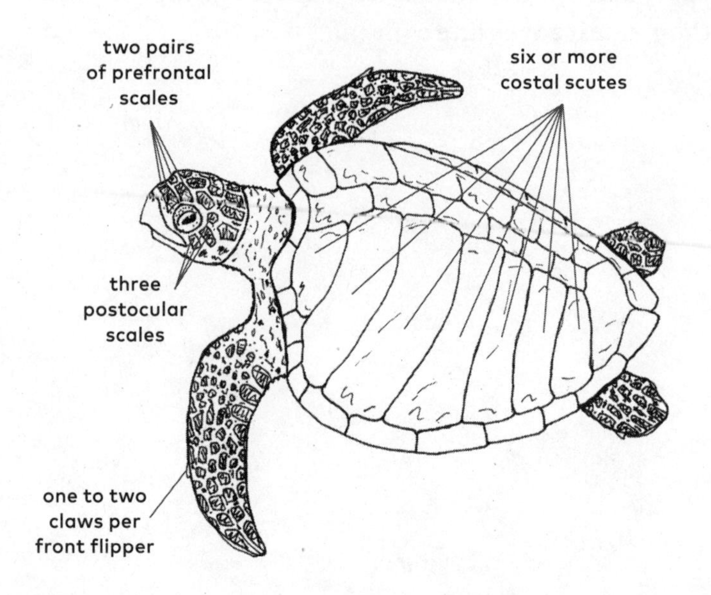

IUCN STATUS: VU—Vulnerable

CARAPACE LENGTH: 60–75 cm (24–30 in.)

WEIGHT: 35–45 kg (77–99 lb.)

MAIN DIET: Crustaceans, mollusks, fish, algae (opportunistic omnivores)

AGE AT MATURITY: 11–16 years

REMIGRATION INTERVAL (between nesting seasons for a given female turtle): 1 year

INTER-NESTING INTERVAL (between nesting events in the same season): 25 days for arribada females, 17 days for solitary females

NUMBER OF CLUTCHES PER NESTING SEASON: 2–3

NUMBER OF EGGS PER CLUTCH: 110

INCUBATION TIME: 45–60 days

GOOD TO KNOW: Synchronized mass nesting (arribadas); also nest individually; heart-shaped carapace with a slight hump; beak like a parrot

– 1 –

FROM THE NEST OUT INTO THE WORLD

DESPITE THE EARLY MORNING HOUR, the Costa Rican sun is already high in the sky and burning hot on my head while I shift shovelful after shovelful of black sand. Giant drops of sweat are running down my face and into my eyes. I am the on-site biologist leading a leatherback conservation project called Baulas y Negras located within the Ostional National Wildlife Refuge, and my colleagues and I are digging a deep pit. We took off our T-shirts a couple of hours ago to avoid heat stroke and are now covered in sweat as we stand up to our waists in a trench about 7 meters long and 5 meters wide (23 by 16 feet). Over the past few days of work, aches in various parts of my back have melded into a single widespread pain that makes me wince every time I move. Blisters on my hands that have recently burst are now oozing blood. I snatched only three hours of sleep after my beach patrol last night, and yet my day is just beginning. When we finally finish digging the trench, we are going to spend the afternoon moving huge piles of heavy, wet sand from the water's edge, either packed into sacks that we will carry on our backs or piled into wheelbarrows that we will roll laboriously up the beach. Then we are going to sieve the wet sand and shovel it into the enormous hole we have dug. This is grueling work. So, what's the point?

All seven existing species of sea turtles are threatened with extinction. Every single one is on the International Union for Conservation of Nature (IUCN) Red List of Threatened Species. Hawksbill turtles and Kemp's ridley turtles are classified as critically endangered around the world; green turtles are classified as endangered; leatherback, olive ridley, and loggerhead turtles are classified as vulnerable; and we do not have enough data to classify the status of flatback turtles. For a few species the status of individual populations is far more concerning than the global figures suggest. Many have suffered a drastic decline in the past fifty years. In fact, things look so dire for some individual populations and some entire species that they are in imminent danger of being lost forever. This is the case for the Eastern Pacific population of leatherback turtles, the focus of our work in Ostional. Their numbers have decreased so rapidly that at the time of writing it is estimated only about three hundred females nest on beaches from Mexico to Colombia every year. Scientists predict that within a decade there will no longer be any leatherback turtles in the Eastern Tropical Pacific—unless people step up to help them.

And that is exactly what my colleagues and I were doing as we dug our trenches that hot and sticky day. At that point only nine females were still nesting in Ostional; however, this number would continue to drop in the coming years.

The main goal for many sea turtle conservation projects worldwide is to produce and protect as many sea turtle babies as possible to ensure a new generation of turtles makes it into the world. In practical terms, this means that we patrol the beach every night so females can lay their eggs in the sand undisturbed—or sometimes straight into a bag that we hold under their tails. When we resort to bags, we take the eggs to a safe place to incubate and hatch.

Why do we do this? As I'll explain in detail later, on many of our nesting beaches, sea turtle eggs are stolen by people—by poachers or nest robbers (*hueveros* in Spanish). In addition, there are many natural predators only too happy to fill their stomachs with sea turtle eggs, including raccoons, coatis, skunks, vultures, and crabs. Unfortunately, predator numbers have increased considerably in the past few years. As natural habitats are destroyed, we are losing increasing numbers of large predators that keep the populations of smaller predators like raccoons under control. Additional predation comes from pet dogs that spend most of their time roaming freely, allowing them to dig up nests and gorge themselves on eggs.

But that's not all. Higher tides and rising sea levels driven by climate change are causing constant erosion of sand from beaches. In other words, grain by grain, sand is being carried away. Sea water spills into nests, and eggs can end up completely submerged. Within just a couple of weeks whole sections of beach can be washed away, along with nests full of eggs that were incubating in the sand.

To protect eggs after they have been laid, many organizations rely on guarded hatcheries (*viveros*). The goal is to guarantee that the eggs incubate successfully. We, too, hope to protect as many sea turtle eggs as possible until they hatch. And that is why almost every one of our nesting seasons starts with a scenario like this: We fill a trench in the beach with clean sand and erect fences for protection and shelters for shade so the eggs in the nests can incubate in peace—guarded twenty-four hours a day, seven days a week by project staff and volunteers.

Ideally, the weeks-long, sweat-inducing work to build these hatcheries is completed before the first females arrive to lay their eggs. Unfortunately, last night in Ostional a female got the jump on us and we are now running out of time because the leatherback female, following the inter-nesting intervals of her species, will come back in ten days to lay her next round of eggs. We must complete our hatchery before she returns.

The most arduous and most boring parts of building the hatchery are digging the pit and sieving the sand to fill it. The hatchery is protected by a two-layered fence made from anti-aphid mesh and chicken wire. The fence is buried up to 1.2 meters (4 feet) deep and stands 1.5 meters (5 feet) tall. It surrounds the whole area to keep it safe and secure. At least half the hatchery is also shaded by an awning made of bamboo poles and camouflage netting. Inside the hatchery we use string to mark off 1-by-1-meter (1-by-1-yard) squares on the sand to guide us as we rebury clutches of eggs in a checkerboard pattern to allow the nests in the small enclosed area to be more closely spaced than they would be in the wild.

We usually work for a few hours pre-breakfast before the sun gets too hot, and then a few hours in the afternoon after the sizzling heat of midday has passed. At night, we patrol the beach. Some of the projects I've led over the years have had not one but two hatcheries, with areas two or three times as large as the small one we are working on. These pre-season preparations are definitely not my favorite part of the job but are necessary to ensure the survival of the eggs and the hatching of as many baby turtles as possible.

Even when a natural nest incubates successfully, it's exceedingly rare for 100 percent of the eggs to hatch; hatching success rate is on average closer to 50 or 60 percent. We can raise this to an average of 70 to 90 percent by incubating the eggs in a hatchery and taking additional precautions. These precautions include using cylindrical cages of rabbit fencing covered with insect netting, which we install over every nest. Each cage is buried 20 to 30 centimeters (8–12 inches) deep to keep crabs from burrowing their way into the nests and eating the eggs. The netting prevents flies from laying their eggs on top of the nests, where the larvae then worm their way down into the egg chamber and eat the eggs and baby turtles inside.

Of course, the cages and fencing also prevent the hatchlings from scrambling directly to the ocean. This means we must watch the hatchery day in, day out and check each nest every twenty minutes. We remove recently hatched babies from the cages right away and transfer them to a dark cooler filled with moist sand. During the day, we wait until nightfall or until it rains so temperatures will be lower before we release the hatchlings. At night, we first gather them into a group and then release them all at the same time.

To estimate our hatching success rates—and therefore the success of our interventions on the turtles' behalf—we dig up nests after the eggs have hatched and inventory the contents so we can see how many babies hatched successfully and how many did not. We also open any remaining eggs to figure out why they didn't hatch. Did an embryo begin to develop or did a bacterial or fungal infection prevent that from happening? This important information helps us judge the effectiveness of our conservation strategies so we can improve them if we need to. Nest excavations are, of course, anything but pleasant. Even though we open the nest within twenty-four to forty-eight hours after the main group of hatchlings has left, many of the remaining eggs are already half rotten. Little wonder I associate nest excavations with an awful stink that sticks to hands and clothes despite the latex gloves we wear. When I was working on my master's, I sometimes excavated up to twenty nests a day. To neutralize the stench, I often tied a cloth sprayed with perfume over my mouth. Unfortunately, it didn't help much and spoiled the fragrance of my favorite perfume for me forever.

On many beaches we have to artificially shade the nests so they don't overheat, as too much heat inside the egg chamber would cook the eggs. Temperature is an important factor in the development of sea turtle babies. Temperatures that are too low (under 24 degrees Celsius/75.2 degrees Fahrenheit) are just as deadly for developing embryos as temperatures that are too high (above 36°C/96.8°F), especially if they last for extended periods. In many places, beaches have been developed and natural vegetation that provides shade has been removed. In addition, climate change is causing global temperatures to rise. This means the temperatures on many of our nesting beaches are beyond lethal limits. Luckily, we can use canopies to lower temperatures enough that the eggs have no problem incubating. This allows us to produce thousands of tiny turtle babies every season, and we can supervise their scramble to the ocean.

Incubation temperature plays another important role in the hatchlings' development. Unlike human babies and the offspring of many other animals, the biological sex of sea turtles is not determined by sex chromosomes but by the temperature of the sand in the second third of the incubation period, a thermosensitive time during which the prevailing temperatures influence the embryos' development. When temperatures are higher, more females develop; when they are lower, more males. The phrase "hot chicks, cool dudes" is a good way to remember this. Temperature-dependent sex determination, or TSD, is known from only a few species of reptiles and fish. The details of the process are not yet fully understood, but one of the enzymes that appears to be at play here—aromatase—becomes more active as temperatures rise. It converts the male sex hormone testosterone into the female sex hormone estrogen. The temperature difference between a nest with 100 percent males and one with 100 percent females is only about 2.5°C (4.8°F). There is, however, a key temperature—called the pivotal temperature—at which the embryos develop into 50 percent males and 50 percent females. For most species of sea turtles, this is about 29°C (84°F).

Climate change and the resulting warming of the Earth has meant that, in general, more females than males have been hatching for the past few decades—often at drastically skewed ratios as high as nine to one. This development concerns me greatly, because even though we might not be seeing the effects right now, the lack of males in future years will present yet another threat to the already endangered populations of sea turtles.

A FEW YEARS LATER we are standing once again on that same beach in Ostional. The hatchery we are building this time is intended not for our own conservation project but for a joint BBC/CBC television documentary about the inner workings of an olive ridley turtle nest. Researchers still have much to learn about what happens inside a sea turtle nest. Our options for observing and examining the eggs and the embryos developing inside them are limited, and so the months of incubation are still shrouded in mystery. The documentary filmmakers want to shed light on some of these secrets. Because of our expertise in building hatcheries, as well as the large number of sea turtles that visit Ostional, the television team is here to capture never-before-seen glimpses into an incubating sea turtle nest on camera.

Our beach is famous for a phenomenon called *arribada*, which is Spanish for "arrival." Almost every month—normally during the week before the new moon—female olive ridley turtles arrive by the thousands in synchronized mass landings to lay their eggs. For three to ten nights in succession, up to half a million females all nest at the same time. Imagine being on a one- to two-kilometer stretch of barely moonlit black-sand beach, when, suddenly, as though someone has blown an inaudible whistle, thousands of sea turtles start crawling up the beach, side by side and even on top of one another. A steady stream of sea turtle silhouettes flows from the water and back until female turtles cover every inch of sand. Some are digging their nests while others are already laying their eggs. Eggs—both intact and broken—and sand fly through the air and can hit you right in the face if you're not careful. You can't go more than a couple of feet without almost stepping on a turtle. It's not unusual for a female to get caught in the shoulder strap of a carelessly placed backpack and drag it across the sand. We've lost equipment in the sand or sea more than once that way. All around, you can hear the thudding sound of sea turtles using the bony part of their bottom shell, called the plastron, to compact the sand on top of their eggs. Only a few females come to lay their eggs during the day while the sun burns down relentlessly, but every night when the sun sets and the tide comes in, the spectacle begins anew, and it lasts until the first rays of sunlight hit the beach in the morning.

This extraordinary natural spectacle—which we still know very little about and is simply breathtaking to experience—has been featured in many documentary films and glossy magazines. Anyone who has seen *Blue Planet II* will be familiar with the impressive drone footage taken of just such an arribada in Ostional. These mass nesting events can be observed at barely a dozen beaches around the world, most of them in Central America and two in India. Ostional is one of two beaches where the largest arribadas have been documented.

Over the next few weeks, we hope to transfer a total of five clutches of olive ridley turtle eggs to the sand nests we've built at the hatchery. There, the eggs can be filmed through sheets of plexiglass, and the temperature in each nest can be monitored from the outside using a sophisticated temperature sensor. We will also use microphones to record sounds coming from the nests. Scientist and sea turtle nerd that I am, I find this incredibly exciting. A group of female (!) scientists will investigate different aspects of the sea turtle nest on camera. A couple of male scientific presenters round out the team; their job is to provide an engaging commentary on the proceedings for the television audience.

Finally, it's time. The hatchery is ready and the film crew has arrived. Their stay has been timed to give them a good chance to film an arribada and for us to collect the five clutches of eggs that we need during the mass nesting event. But, as so often happens, things do not go as planned. After waiting in vain for several nights for an arribada, we decide to gather eggs from solitary nesting females. Luckily we have enough of these in Ostional. This will give the eggs plenty of time to incubate, and we won't have to delay the second scheduled slot for filming in forty-five days, a time we chose because that would be the earliest the eggs would hatch after being transferred to the artificial nests.

And so, the next night, we set off to find these females. I walk up and down the waterline, and finally, after a couple of hours, I spot a dark "stone" with a distinctive dome shape gliding out of the water. Right away, I flash a signal to the film crew waiting a few miles away, causing mild chaos in the drowsy group as everyone tries to get themselves and their equipment organized as quickly as possible. Finally, the convoy sets off and hurries along the beach toward my red light. Within fifteen minutes, the female producer, the two female directors, the two cameramen, the sound assistant, the two scientific presenters, and three of our local workers, slightly out of breath, reach me and the turtle and get ready to film the process of collecting the eggs.

Ideally, we catch the eggs in a specially designed bag as the turtle mother is pressing them out of her cloaca and before they drop into her sandy nest. To do this, the person collecting the eggs lies on their stomach behind the turtle and carefully holds the bag under the cloaca. The bag must be extremely sturdy and large enough to accommodate all the eggs because a single clutch can sometimes weigh several kilograms (10 pounds or more) and, depending on the species, can include up to two hundred eggs. How do these fragile-looking eggs survive a drop of 45 to 75 centimeters (18–30 inches) without sustaining any damage? Sea turtle eggs, just like other reptile eggs, have a flexible, parchment-like shell that cushions their landing. The shell is also permeable to water and gases, so the embryo has a constant supply of oxygen as this gas diffuses unobstructed through the shell from the sand. The shell itself consists of a fibrous organic inner membrane and an inorganic or mineral outer layer of calcium carbonate, which forms over the organic membrane as crystals of aragonite or calcite. This amazing packaging creates a safe space for the embryos to develop.

As soon as the female has dropped all her eggs, we carefully remove the bag from the nest. If this is done skillfully enough, the turtle is none the wiser and begins to cover her nest with sand thinking her eggs are still inside. This avoids stressing the turtle unnecessarily. If you fumble around, make contact with the cloaca, or generally are too clumsy, the female might take off midway through laying her eggs. Clearly, this is a scenario we want to avoid at all costs. And that's why we practice collecting eggs with new workers using turtles sculpted out of sand and Ping-Pong balls, explaining in detail what they need to look out for. Our film crew, including the commentators and scientists, also got this training.

After all the eggs are in the bag, we carry them carefully to the hatchery, jostling them as little as possible, because the embryos begin to develop the moment the eggs fall from the cloaca and are exposed to oxygen. This triggers a continuation of the process of cell division that began in the mother's body after her egg cells fused with the male sperm cells but then stalled in the oxygen-depleted environment of her oviduct. After the eggs are laid, tiny turtle embryos—nourished by egg yolk—start to form from these clusters of cells. Ideally, we want to complete the relocation of eggs within the first six to twelve hours of their being laid, because after that the embryos inside attach themselves to the top of the shell as they begin to develop. If the orientation of the egg changes after this attachment—for example, if it's turned upside down—the embryo can suffocate. It's also important for us not to hold the bags of eggs too close to our bodies as we carry them, because the human body temperature in the range of 36.5°C (97.7°F) is far too warm for the developing embryos. Therefore, we hold the bags away from our bodies with our arms stretched out to the side, keeping the bags as still as possible. This also ensures we do not land on top of the eggs if we stumble and fall. As you can imagine, it's tough to hold this position when you might be walking several miles.

The time it takes for an embryo to develop into a fully formed baby turtle depends on the temperature of the sand during the incubation period. If the sand is on the cool side, say around 25°C (77°F), because the nest is in the shade or because it has rained for an extended period, incubation can last from sixty-five to seventy-five days. If, however, it's in direct sun and there was hardly any rain, higher temperatures of up to 35°C (95°F) cause the embryos to develop more quickly and the eggs might hatch after only forty-five days. Babies that hatch after a shorter incubation period are often smaller, and babies that hatch after a really long time in the egg tend to be larger and also faster. This happens because it takes a while to transform the individual building blocks supplied by the yolk sac into a turtle. Shorter times only allow for the development of a small turtle, although with ample yolk reserves for energy; longer times allow for the construction of a larger and faster baby. The factors that are most important here are the baby's self-righting ability, meaning its ability to right itself after flipping over, and its crawl and swim speed. The ability to right itself within the first two hours of leaving the nest, crawl over the sand as quickly as possible, and then swim speedily raises the chances that a baby sea turtle will survive its transit from the nest to the water and then swiftly get beyond the reach of aquatic predators.

Olive ridley turtles, thank goodness, lay an average of only about one hundred eggs the size of Ping-Pong balls with a combined weight of about 3 kilograms (6–7 pounds)—so carrying them to the hatchery is not as difficult as carrying, for example, a clutch of leatherback turtle eggs. In general, smaller species lay more and smaller eggs, and their offspring are smaller than those of larger species. The Australian flatback turtle is the exception. Although it's one-third the size of a leatherback, its eggs are almost the size of a leatherback's eggs. Interestingly, the size of individual females has little to do with the size of their eggs—bigger females lay more eggs but not necessarily larger ones.

The reason for this is that producing larger eggs, and therefore larger offspring, comes at the cost of producing fewer offspring. In evolutionary theory, the adaptive value—or Darwinian fitness—of a hereditary trait is determined by its effect on the number of offspring. In other words, an adaptation is better if it increases the overall number of offspring or the chances of survival for individual offspring. Therefore, under exactly the same environmental conditions, an individual with higher evolutionary fitness has more surviving offspring than one with lower evolutionary fitness. However, one adaptation is not always fundamentally better than another—as in the case of sea turtles. More eggs mean more offspring, but larger eggs lead to larger offspring with higher survival rates. The point at which an equilibrium is reached between these two desirable but mutually exclusive situations is explained by the theory of optimal egg size, which states that natural selection has optimized egg size at the point where the increased evolutionary fitness associated with larger egg size equals the reduction in evolutionary fitness caused by a lower number of surviving offspring. Or, to put it more simply: the goal is to produce the largest eggs you can while still producing as many offspring as possible.

Sea turtles generally aim for quantity rather than quality. Instead of producing a few young that the mother looks after with selfless devotion, females lay hundreds of eggs each season in multiple nests that they spread over different sections of beach or even over different beaches. Other reptiles also use this strategy. But unlike, for example, mother crocodiles, who watch over their nests and help their newly hatched young reach the water, mother sea turtles disappear into the sea after laying each clutch of eggs. As a result, of course, many clutches of sea turtle eggs never even get to hatch but instead are plundered by human or animal nest robbers or washed away by the ocean.

But that's why we are here. When we arrive at the hatchery, we dig a nest the size and shape of the nest a mother olive ridley turtle would build. It needs to be freshly dug because otherwise the sand dries out, and dry sand will destroy the delicate environment of the eggshell and alter its permeability, damaging the embryo growing inside it. Given the importance of temperature for incubation, it's essential to get the correct depth as well as the correct dimensions for the egg chamber. Both depend on the size, or rather the flipper length, of each species: larger species dig deeper nests than smaller species. As olive ridley turtles are one of the smallest species of sea turtles, their nests are only 45 centimeters (18 inches) deep, whereas the giant leatherbacks dig the deepest nests with an average depth of 75 centimeters (30 inches). Leatherbacks are also the only sea turtles to dig nests in the shape of a boot; the nests of other species look more like a traditional light bulb with the bulbous part at the bottom. In all cases, the egg chamber is the deepest part of the nest and must be large enough to accommodate all the eggs in a clutch. Although the mother turtle seals the entrance to the egg chamber with sand, there is still room inside the chamber for gas exchange and, later, for the freshly hatched babies to move around.

Our challenge for the documentary film project is to dig the nest in a way that allows a direct view into the egg chamber through the plexiglass. But we manage to do that and finally, using gloves, we carefully place each egg individually into the nest we have prepared. We also place a temperature sensor among the eggs and connect it to a thermometer outside the nest. Finally, we cover the nest with sand, just as a mother turtle would do. Over the course of two nights, we collect the four additional clutches of eggs needed for filming and transfer them all, just like the first one, into our hatchery. For the next several weeks, we will observe the incubating eggs through the panels of plexiglass we have installed in the nests, and film the eggs while we monitor the temperatures in the nests using our sensors.

Meanwhile, we have not yet given up hope that we might get to film an arribada. The time allotted for filming has even been extended by a week. One night the project staff and the television crew are all sitting on the beach waiting for another turtle to emerge and nest, so that I can attach a satellite transmitter to her shell. While we are keeping watch and the crew is snoozing, the sand beneath us and around us suddenly takes on a life of its own. Thousands of nests have chosen this night of all nights to hatch. As most eggs in Ostional are laid within a few days of each other every month, the little sea turtles usually all hatch within a few days of each other. Millions of young olive ridley turtles are wriggling their way out of the sand almost simultaneously and racing for the sea. At times like this, it is almost impossible to walk along the beach without fear of stepping on a baby turtle. The camera crew can barely contain their excitement as they eagerly begin filming. One of the presenters, a gray-haired and usually extremely dignified Brit, is crawling on his knees in the sand in his light-colored khaki pants repeatedly uttering squeals of joy, his nose almost touching the ground so he can get a better view of the tiny turtles in the glow of our red lights, which do little to brighten the darkness. What a delight for everyone after the gloomy mood over the past few days in the absence of an arribada.

Unfortunately, the nature spirits are not on our side and most of the film crew leaves without setting eyes on an arribada. It's now almost Christmas and further delay is out of the question because everyone wants to get back to their families. One cameraman is left behind to hold the fort over the holidays. And wouldn't you know, on Christmas Eve of all days, the inaudible starting signal for the synchronized mass nesting is given, and thousands of olive ridley turtles appear on the beach in Ostional. For five days, the cameraman runs around like a headless chicken filming day and night and sleeping barely if at all. Then this part of the turtles' life cycle is also captured on film, and while the film crew sifts through the raw footage, we can focus on the incubating nests.

THE FILM CREW RETURNS forty days after we transferred the eggs to the hatchery. This time, they want to document the eggs hatching in the nest. The temperature in the nest has been perfect all along, steadily climbing as the eggs incubate. This is happening not only because outside temperatures are rising with the onset of the dry season, but also because of the warmth created by the increased metabolic activity of the developing turtles. This is generally a good sign indicating things are happening inside the eggs. We can't yet detect any activity through the plexiglass. The eggs lie still and round in their cradle of sand. But we do have a scientist who specializes in the vocalizations of baby sea turtles. She slips an omnidirectional microphone into the nest through holes in the plexiglass so she can eavesdrop on the embryos.

Most people think that turtles don't make any sounds and also don't hear very well. However, scientists have known for quite some time that turtles are pretty good at picking up sounds in frequencies between 100 and 1,000 hertz, with the greatest sensitivity at about 300 hertz. Human speech frequencies have a median pitch at around 125 hertz for male voices and a median pitch at around 220 hertz for female voices. This means sea turtles have no difficulty hearing our conversations but likely aren't able to hear the high-pitched cries of a human baby. We assume that the sounds sea turtles hear on land are slightly muffled as their hearing is adapted to the density of water.

Various recent studies have investigated the social behavior of aquatic turtles, both those that live in fresh water and those that live in salt water. For decades, many scientists have been puzzling over the mysteries of how hatching happens and how mature babies within the eggs somehow manage to coordinate so that almost all of them leave their eggs at the same time and make their way to the surface as a group. This process is known as protocooperation. It's a form of ecological interaction in which all participating organisms benefit, similar to a symbiotic relationship. In the case of sea turtles, you could also call it a rudimentary form of social behavior. For one thing, this cooperative behavior benefits baby sea turtles because they don't have to undertake the difficult task of climbing out of the egg chamber on their own; many pairs of flippers help them do this. Furthermore, being in a group gives them added protection. Predators are overwhelmed by the number of babies and snap up just a few of them while the rest reach the water safely.

Until recently, we had no idea how this basic form of cooperation worked in detail and how the babies synchronized their hatching. Then, in 2012, a study was published that documented the vocalizations of a freshwater turtle that lives in the Amazon, the giant South American river turtle, for the very first time. The authors suspect that the baby turtles use these vocalizations to coordinate their hatching. A couple of years later, the co-author of the study presented the findings at an international symposium for sea turtle research. I was in the audience and sat spellbound, hardly believing my ears as I listened to the chirping, clicking, mewing, and clucking over the loudspeakers. A lively mix of vocalizations from a species of turtle we had all assumed made no sounds at all! After this presentation, it was clearly time to listen to sea turtles as well.

One of the scientists who listens to turtles is my colleague Lindsay McKenna Jackson, the scientist who is now adjusting her microphone in the nest. For days, we hear nothing. Then, suddenly, the microphone picks up distinct, pulsing chirps. This is the first time I have ever heard sea turtles vocalizing, and I'm so excited I hold my breath so as not to miss anything. And a few days later, the first babies hatch.

From the other side of the plexiglass, we experience the hatching process up close for the first time. I am transfixed by the spectacle. First, we see movement within the eggs. Then, suddenly, small, pointed triangles peek out from some of them. These are the baby turtles' egg teeth, which they use to cut open their shells so they can poke their heads out. Over the next few hours more and more tiny turtle heads appear, gradually followed by their bodies, which slowly but surely wriggle out of the eggs. Many widen the holes with the elbows of their front flippers and then stick their flippers out. From there, they squeeze the rest of their bodies bit by tiny bit through the hole. It's a tiring process, and every once in a while the babies need to rest. But eventually almost all are completely hatched and lying exhausted on top of their broken eggshells.

Over the next few hours, any remaining yolk in the yolk sac is absorbed into their abdominal cavities and the hole in their abdominal walls dries out to close it off. The yolk will provide food and energy to the young turtles in the coming days and weeks, fueling their arduous journey from the nest to the water and from there through the strong currents out into the open ocean.

This is why it is harmful when baby turtles are kept for days in coolers or, worse, in pools, just so tourists can release them on the beach for a dollar, a practice in many tourist spots, particularly in Asia. In addition to burning through their precious stores of energy prematurely, the animals are also denied the protection of being in a group. These attractions are offered under the guise of sea turtle conservation, and this is just one of many reasons why you should not support them. It's quite different if you are allowed to be present when a whole nest is released into the ocean at night or on a rainy day or at dusk right after the eggs have hatched.

Hatching is not the end of the babies' ordeal, because they must now make their way up to the surface of the sand. This can take from three to five days, depending on the species of turtle and the depth of the nest. Many people imagine the babies painstakingly digging their way up through the sand. What actually happens is that the young turtles benefit from the architecture of the nest. The egg chamber now contains about one-third more open space after the fluids in the eggs seeped into the sand as the babies hatched, and the shells are now lying on the floor of the egg chamber, flattened by hundreds of tiny flippers. The extra space causes the sand above to collapse and slowly trickle down into the egg chamber. As soon as the babies move their flippers, sand falls down and gathers in a pile underneath them. And then more and more and more. As the sand slowly trickles down, the little turtles are gradually carried closer to the surface; it's almost like being in an elevator.

They have no idea, of course, at what time of day or night they will arrive at the surface. But an ingenious natural process ensures that they will not make their way to the water in the hottest hours of the day, when they would literally roast on the sand. High temperatures paralyze them before they reach the surface, which means that many babies finally crawl out of the nest shortly after the temperature drops significantly, as it does, for example, right after sunset or at the start of a heavy rain shower after a few hours of sunshine. Then hundreds or even thousands of babies suddenly emerge from the sand at the same time and run for their lives toward the water.

The hatchlings in our nests are still busy absorbing the remains of their yolks. As we watch them, the first sand from the surface collapses, gradually blocking our view through the plexiglass. We lose visual contact and will have to wait until the babies reach the surface to see them again. But for days, nothing happens. We can see, via our sensors, that the temperature in the nests is rising—toward potentially lethal levels. The film crew is getting nervous and worried. My colleagues and I try to reassure them that this rise in temperature is completely normal. But in the end, they lose their nerve and open one of the nests. Of course, everything is just fine. Our babies had almost made it to the surface and are raring to go. After this discovery, we leave the other four nests to their own devices and wait.

Generally speaking, you should always leave turtles to hatch on their own and not handle them unnecessarily. The effort of hatching and getting from the bottom of the nest to the surface forces the babies to use their muscles and completely expand their lungs. This gets everything working properly, much like the first cries of a human baby. Too much assistance could be counterproductive and harm the babies' chances of survival as each of the natural exertions in the nest prepares them for the dangerous world outside. The best way to help is to clear a path from the nest by removing driftwood, smoothing out larger dips and footprints, filling in crab holes, and then simply being there to drive away birds. You should pick the babies up only in an emergency and then only under supervision and with gloves to protect the hatchlings from all the bacteria and chemicals on your skin (creams, sunscreen, bug sprays, nicotine from cigarettes).

The babies also need to imprint on the beach so the females among them can return later as mature adults to lay their eggs. We don't yet know exactly how this imprinting works, but we assume that the journey to the water is a crucial part of the process. Therefore, we don't want to deprive the animals of any points of reference they need to help them find their way back as adult turtles. We don't want them to end up swimming around lost in the ocean. The Kemp's Ridley Headstart Project in Texas was also mindful of this in their work. Starting in 1978, the project incubated all its Kemp's ridley eggs in temperature-controlled climate chambers to save the species. In hopes of giving them a better start in life and increasing their chances of survival, the freshly hatched babies were raised in ponds for months until they grew too large for most predators to swallow. Only then were the turtles released into the Gulf of Mexico. Unfortunately, at the beginning of the project, nothing was known about the imprinting of the babies on their natal beaches. As soon as scientists learned about this, the project changed its protocol and the babies were allowed to crawl down the beach immediately after they hatched. Then volunteers caught them using nets and took them to the rearing ponds. That way, the project ensured that the females, as well as the males, had a clear destination when the time came to mate and lay eggs.

The babies confined in our hatchery can't run directly to the water either. After all the eggs finally hatch, we carry the turtles to a designated spot high up on the beach where the sand gives way to vegetation and release them there. We make sure no crabs or birds eat them as they dash to the water as a group. After two months in their sandy womb, our little sea turtles finally reach the water, with no parents to protect them but in the company of their siblings, and now begin their long life journey.

HAWKSBILL TURTLE

CR

Eretmochelys imbricata

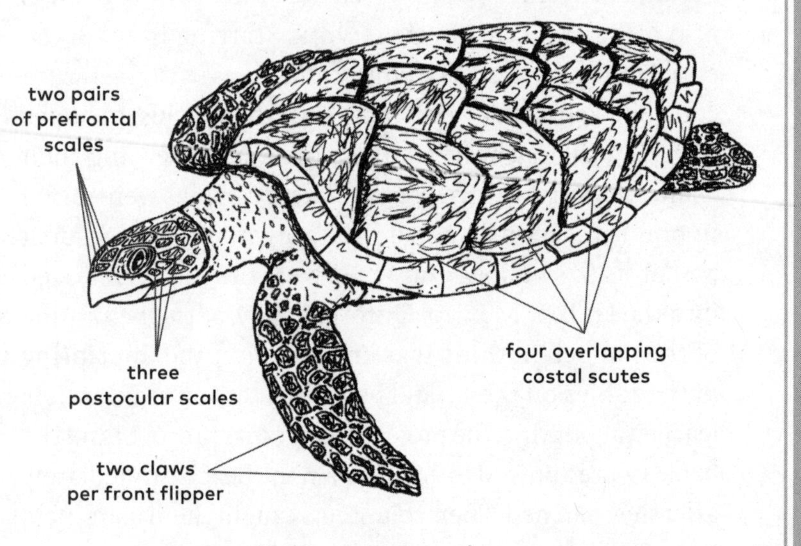

IUCN STATUS: CR—Critically Endangered

CARAPACE LENGTH: 75–90 cm (30–35 in.)

WEIGHT: 45–70 kg (99–154 lb.)

MAIN DIET: Sea sponges and soft invertebrates

AGE AT MATURITY: 20–25 years

REMIGRATION INTERVAL: 2–3 years

INTER-NESTING INTERVAL: 14 days

NUMBER OF CLUTCHES PER NESTING SEASON: 3–4

NUMBER OF EGGS PER CLUTCH: 150

INCUBATION TIME: 55–65 days

GOOD TO KNOW: Shell is known as tortoiseshell and highly prized; beak looks like a hawk's beak, hence the common name

– 2 –

DIVING IN A SEA TURTLE KINDERGARTEN

I'M IN A BUOYANT MOOD as I walk down the long boardwalk to the water. The Red Sea, sparkling in the first sunlight of the day, beckons. I have the place to myself. Even the dive station, which is normally bustling with activity, is quiet. Only the poor donkey that pulls the cart loaded with oxygen tanks down to the water for customers is awake and waiting for his morning feed. When I arrive, I sit down on the edge of the jetty, look out over the almost mirror-like surface of the water, and take a deep breath. The sun is just rising above the horizon and warming the planks. I pull on my flippers, brush the hair out of my eyes, and put on my diving mask. I attach the mesh bag containing my research tools to my weight belt. Slowly, I lower myself from the jetty and slip into the tranquil sea. The water feels pleasantly warm against my skin and washes away the last traces of sleep. I briefly check the fit of my diving mask and snorkel and then swim with strong, calm strokes of my flippers along the edge of the reef.

The rays of the early morning sun penetrate the surface, creating an ever-changing pattern of shiny ripples on the seafloor and corals. Thousands of fish teem in the water around me, shrimp use their antennae to tentatively feel their way out of cracks and crevices, Christmas tree worms emerge from their burrows and unfurl their crowns, and sea stars push themselves bit by bit across the sandy bottom. The day on the reef began shortly before sunrise, and everyone is getting ready for their daily routine. The cleaner wrasses at their cleaning stations have their fins full servicing their clients quickly enough, as a number of large fish that will later swim out to deeper water to look for food are still moving around the reef waiting for the wrasses' attention. I take a deep breath in through my snorkel and dive deeper into this fantastic world. A school of small colorful fish floating over a coral attracts my attention. In thrall to an invisible current, the school sways back and forth like a dancer as it maintains its position. I surface, breathe deeply, and swim on.

This is back in my early days as a student, and I'm here in Egypt in 2005 for a university project. Every morning starts like this. I soon come to a cleaning station marked by a white stone with a black number 8 painted on it. I pull an underwater writing tablet, a pencil, and a ruler out of my mesh bag. I place the ruler carefully next to the cleaning station and take up my observation post directly above the station. Two cleaner wrasses have teamed up to offer their services here. I use my tablet to record the size of the clients and how much time the cleaner fish spend with each one. Later I will compare these data—together with data I have gathered from other cleaning stations as part of my first university research project—with data from cleaning stations where only one cleaner wrasse is at work. I have marked a total of twenty cleaning stations scattered across the reef and I visit them twice a day, shortly after sunrise and shortly before sunset, when the fish on the reef are most active.

As I focus on my lively pair of cleaner wrasses, I catch sight of a movement out of the corner of my eye. When I turn around to investigate, I see a sea turtle floating near the surface right above a steep cliff of coral next to me. It's digging through the coral with its sharp beak looking for food. I carefully swim closer. Completely unimpressed by my presence, the turtle continues to probe the gaps in the coral and finally finds a sponge, which it expertly extracts and eats, savoring every morsel. I can hardly believe my luck. This is the first time I've seen a sea turtle in the wild. Up until now all I know about them has come from television programs, books, and my visits to the Aquazoo in Düsseldorf. Fascinated, I soak up every detail of this encounter.

It has a shell, just like a tortoise or freshwater turtle. But the shell is flatter, and on this individual the edges are serrated. Instead of feet, it has four flippers, and I can see the tip of a tail under its shell. Its head is small and slender, and its mouth looks like the bill of a hawk. I don't see any ears, but I do see two alert, slightly protruding eyes, which are actively looking for food. Its nostrils are between its eyes. Its body, including its shell, is covered with scales. These unique morphological features are characteristic of reptiles, a group that includes turtles, along with lizards, snakes, and crocodiles. The horny scales on the sea turtle's shell are called scutes, or *scuta* in Latin—the plural of *scutum*, which is the name for the shield carried by Roman legionaries.

Both the softer scales on the sea turtles' skin and the scutes on its carapace (top shell) are made of keratin, a material similar to our fingernails and hair, although our alpha-keratin is softer than the beta-keratin of reptiles and birds. I notice two claws on each front flipper of "my" sea turtle and one on each of its hind flippers. That, I will learn, is an important feature you can use to tell the species apart. Although I don't know it yet, I am looking at a hawksbill turtle foraging for food.

As I watch, I wonder how old it might be. That's difficult to tell, because it is virtually impossible to estimate the age of a sea turtle just by looking at it. Their scutes don't have any visible growth rings and you can't look into their mouths to estimate their age based on the wear of their teeth, as you can with dogs and horses, because turtles don't have any teeth. Instead, the upper and lower jaws are covered with keratin sheaths to form a beak. The only way we know how to more or less accurately estimate the age of a sea turtle is by looking at the humerus (the upper-arm bone of the front flipper) and using skeletochronology to count the internal growth rings much as you would count the growth rings in a tree. To do this, the turtle must be dead—so it's not a good system for estimating the age of live animals. Instead, scientists divide sea turtles into size classes based on the length of their carapace. This allows us to draw some conclusions about their stage in life, and whether we are dealing with a hatchling, a post-hatchling, a juvenile, a subadult, or a mature sea turtle. None of this is super exact, but it's a useful tool nonetheless. I estimate my turtle to have a carapace length of about 50 centimeters (20 inches). It seems huge to me, but by the time it's sexually mature, its carapace will have grown to an impressive 80 centimeters (31 inches) in length. I'm probably looking at an older juvenile or a subadult, which means I'm diving on a reef being used by the young sea turtles as a kindergarten, a safe place to hang out as the turtle grows large enough to fend for itself in the open ocean.

I am amazed at how precisely the turtle maneuvers its heavily armored body in all directions using its four flippers—to the right and left and forward and backward—how it maintains its position when it discovers something interesting, and how it expertly uses its beak either as a cutting tool or as pincers. It extricates one sponge after another and moves elegantly and adroitly through the water despite its bulk.

Sea turtles' lives begin on land, but they spend most of their time at sea looking for food and are consequently well adapted to life in the ocean. Their keratin-covered carapace is streamlined to facilitate swimming through water, and parts of their bone structure are pared down. Whereas land-based turtles (tortoises) and freshwater turtles are built from massive bony plates, the skeletons of sea turtles are more like scaffolding, making them lighter and more buoyant. Leathery, scaly skin covers elongated finger and toe bones to form large paddle-like flippers. The front flippers are longer than the hind flippers and, in conjunction with the huge pectoral muscles, constitute the turtle's main form of propulsion. The way turtles swim is strongly reminiscent of the way birds fly. Birds beat their wings and turtles move both front flippers up and down simultaneously in powerful strokes.

As for most marine animals, one of the greatest challenges for sea turtles is their salty habitat, which continuously draws water out of their bodies. Their typically reptilian scaly skin helps sea turtles reduce the amount of water they lose through their skin, but even so, water is lost that must be replaced. To do this, sea turtles have two kidney-like glands in their head, located next to their brain, which continuously filter excess salt from their body and excrete it as a thick liquid through modified tear ducts in their eyes. Under water this process is barely noticeable, but on land it looks as though the turtle is weeping, which has led to many myths about nesting turtles.

Astonishingly, many people have never considered that sea turtles have lungs, just like whales and dolphins, and must regularly come to the surface to breathe. If they don't, they could drown or suffocate. In contrast to the automatic breathing of humans, sea turtles consciously control their breathing, which means they decide when they want to surface in order to breathe. This is an eminently practical arrangement when, for example, a turtle is looking for food and doesn't want to interrupt its search. The turtle in front of me seems to think now might be a good time to take a breath, and it lifts its head out of the water. It expels stale air through its nostrils, inhales quickly, and immediately pulls its head back under. Then it resumes its search for sponges.

After it has systematically searched the area and found no more sponges, it swims with calm strokes of its front flippers into the impenetrable blue beyond the reef edge. Mesmerized, I watch it leave until, a heartbeat later, it is gone. It is only now that I realize I have been holding my breath for the whole time as I was floating on the surface and looking down. I now take several deep breaths as I think about what just happened and then, somewhat reluctantly, turn my attention back to my cleaning station.

For the next few hours I can't get my first sea turtle out of my mind, and for once I can hardly wait until breakfast, when I will be able to tell my fellow students all about it. I am one of the lucky few who got the opportunity to participate in the fieldwork portion of a course in tropical marine biology while still an undergraduate—and here I am in Egypt working on a small research project of my own. Most of the other participants are already in grad school, and compared to me they seem so much wiser and more experienced. This is the first time I have ever collected data independently, let alone developed and implemented a research project. After feeling overwhelmed at first, I am now completely in my element. I am beyond happy floating above my cleaning stations for hours on end watching the cleaner wrasses work. The trip to Egypt is all I ever imagined as a child and so much more.

I GREW UP in West Germany in a small town in the Ruhr area, whose defining features were coal mines and a chemical factory. I lived for the vacations my parents regularly took with me and my two younger sisters. As we mostly traveled within Germany, we were more likely to end up at the North Sea than in the Caribbean, but at least we were out of the gray wasteland of the Ruhr.

For as long as I can remember, I wanted two things: to explore life underwater and to move to a country far away. When I was a child, if anyone asked me about my plans for the future, that is exactly what I told them. Most adults, of course, smiled at me indulgently and told me that dreams seldom materialize as life tends to get in the way. Although I didn't completely understand what they meant, I knew they were denying me my dreams—which made me incredibly angry and all the more determined to make my dreams come true. I was certainly not an easy child. I was an awkward mix of mouthy and meek and had an opinion about everything and everyone, which once caused my elementary school teacher to comment that although I thought I knew better than others, I really didn't know what I was talking about. But because I read a lot, I did know a thing or two, and my questions and comments often frustrated those responsible for my education—and probably also exposed the limits of what they knew.

My other passion was books and the adventures I experienced through them. Three titles stand out in my memory: *In unberührte Tiefen* (published in English as *To Unplumbed Depths*) and *Der Hai* (The shark) by Austrian marine biologist Hans Hass, and *In the Shadow of Man* by Jane Goodall. The first two, I swiped from my father's bookshelf; the third, I purchased for fifty cents at a book sale at our local youth library. These autobiographical books opened the door into another world. A real world, because the authors mostly wrote about their own experiences and their exciting lives, filled with travel, adventure, and animals. In my mind's eye, I saw myself romping through the jungles of Africa and diving among coral reefs in the ocean. Being a child, I didn't waste much time wondering about what it might be like to travel alone as a woman or that such expeditions needed to be paid for. My father supported my interests but also thought I was much too fidgety to become a scientist. I couldn't sit still for ten seconds, he told me, let alone for the hours I would need to spend in one spot doing things like observing animals. I think he forgot that I had to sit still to read all those books!

Over the years I learned that my dream career had a name—marine biologist—and that if I went to university and made the most of the opportunities such a degree affords, I could indeed turn my hopes and dreams into a profession. I did everything I possibly could to reach my goal. While my school friends collected articles and posters featuring NSYNC and the Backstreet Boys, I searched all the books and magazines in our school, town, and youth libraries for information about whales and dolphins, which were my first love in those days. I filled multiple folders with the copies I made of magazine articles and chapters from books. Every documentary film about marine mammals broadcast on television I recorded onto VHS tapes, which I labeled meticulously and filed by subject matter. It was no surprise, therefore, that when it came time to choose where to do my mandatory internship as part of my school curriculum, I chose the dolphinarium in Münster. Unfortunately, my school did not allow an internship so far away from my hometown, and so I decided to become an intern on my own time during my summer vacation... and then again over the fall break and the winter break and the following summer. I became a permanent intern.

In the first few years, I was the youngest intern at the dolphinarium, but I wasn't treated any differently. I worked as hard as everyone else and had the same tasks, as well as the same privileges. Whatever our opinions nowadays on keeping dolphins in captivity, it was my first close-up encounter with these fascinating animals, and for my thirteen-year-old self it was a hugely interesting and motivating experience. The care and feeding of the sea lions and dolphins went hand in hand with lively scientific exchanges and conversations about the conservation of aquatic mammals in South America. Not only was there a group of students of animal physiology and behavioral biology from the University of Münster who collected data for their postgraduate research, but there was also the conservation organization Yaqu Pacha e.V. (*yaqu pacha* means "water world" in Quechua), founded by two biologists affiliated with the dolphinaria in Münster and Nuremberg respectively.

I soaked it all up like a sponge. I could bombard everyone with as many questions as I liked and no one seemed to mind: What did I have to do if I wanted to study marine biology? How long do sea lions live? How exactly does echolocation work? Can two different dolphin species communicate with each other? The flood of questions never stopped, and my young brain was bubbling over with new knowledge. It was one of my happiest times in Germany.

In the years that followed, I made other decisions that I hoped would help me later in my studies. For example, I chose Latin as my second foreign language and attended high school in the U.S. for a year to improve my English because English is the official language of science. I must admit that although I was a good student, I wasn't particularly motivated. I simply didn't enjoy most of my subjects. Every morning I grudgingly took myself off to school, where I spent most of the time daydreaming. When I got home, I threw my schoolbag in the corner and didn't open it until I was back in the classroom. I rarely, if ever, did my homework, and I was content with a B instead of studying so I could get an A, which infuriated my teachers and my parents. I was accused not simply of being inattentive but of being positively lazy—although I'm actually not lazy at all. But little of what was on offer at school interested me. Despite that, at the end of my time at school, I had a grade point average that was good enough to allow me to study biology at my chosen university.

During all those years when I sat in my room imagining what my future might look like, it seemed almost impossible that one day I might live far from the Ruhr's oppressively cloudy skies and treeless streets with barely an animal in sight. But I had a plan, and so after I graduated high school, I began my undergrad work in biological sciences at the University of Tübingen. I intended to switch to Hamburg for my master's, where I wanted to specialize in marine biology.

When I traveled to Egypt I was right at the end of my undergrad course work, and I had not enjoyed my time at Tübingen at all. The required courses, which ranged from chemistry to microbiology to botany, were no fun, because most of them were simply rote learning. I began to have serious doubts about my decision to become a biologist. And because I had little interest in most of the subjects, the results of my written examinations were not exactly stellar, and I barely squeaked by in my intermediate exams. I considered abandoning my studies, and my childhood dream seemed to vanish into thin air. Then, out of nowhere, I was offered the opportunity to travel to Egypt and take all the accompanying courses, and that is what brought me to the Red Sea.

BEING HERE IS a complete departure from my previous situation, where I had to squeeze all that useless knowledge into my head. Here, the topics are fascinating and it's all about putting what we learn into practice. We learn how to develop and execute a research project, and how to avoid bias when gathering and evaluating data. On top of all that, I get to work in the most amazing location you could possibly imagine: the Red Sea. As I dive through the colorful corals in this tropical ocean, I see in my mind's eye my childhood heroes Lotte and Hans Hass and Jacques Cousteau swimming along beside me, suddenly showing me their world up close. For the first time in a long while, my own world seems back on track and I stick with my decision to become a marine biologist. The remaining days in Egypt are a dream, and I use the time to forge plans for my future in Hamburg.

Here I am right at the start of my life, and yet I already feel like it's a roller-coaster ride. Sometimes things barely move at all, and it is as though I'm on a slow train to nowhere. Then, suddenly, I'm on a bullet train careering its way through valleys and over chasms, making a brief stop at the top of a mountain before plunging once again into the abyss. Even now, life appears as full of grunt work as it is of adventure and beauty—and it's all happening at once.

Sea turtles must make a similarly difficult journey over the course of their lives. After the tiny baby turtles hatch and dash to the water, they swim out into the strong currents and allow themselves to be swept out to the open ocean. Until recently, we knew frighteningly little about the first years of a sea turtle's life, which is why they are sometimes called the "lost years." Because of their size, small turtles are easy prey and many don't even survive the first few hours after hatching. I once found more than forty-five baby turtles in the stomach of a single mahi-mahi.

A statistic of somewhat obscure origin says that only about one baby in a thousand survives to maturity. Depressing odds.

And that's why a major goal for scientists like me is to find out more about the early life stages of sea turtles: Where exactly are their nurseries—the places the baby sea turtles head for once they leave their nesting beaches—and how long do young sea turtles spend there before moving to their coastal kindergartens? Only when we know this can we do a better job of protecting them. How can we do this? It's relatively easy to attach a satellite tracker to the shell of an adult turtle, but with babies that weigh no more than 10 to 40 grams (0.4 to 1.4 ounces; the equivalent of two to eight nickels or two to eight 20p coins) it is much more of a challenge. Although most trackers are relatively light, they are still too heavy for baby turtles. An added problem is that, in most cases, trackers don't stay attached for long because little turtles grow so quickly. Using stronger adhesives doesn't help; that would restrict the growth of their shells too much, which could lead to deformities.

Despite the difficulties, a few scientists—the most prominent being Dr. Kate Mansfield—have taken on the challenge, repurposing solar-powered mini-trackers designed for birds and finding new ways of attaching them. For one method they first apply a base layer of acrylic varnish—the kind you get in a nail salon—to the shell. Then they attach the tracker on top of the varnish layer using an adhesive mixture made of neoprene and silicone. This way, the trackers stay attached to the young turtles for longer but don't interfere with their growth. New data relayed by these trackers have shown that the babies use aggregations of sargassum (brown seaweed that floats freely in major ocean currents) as places to hide. The seaweed also provides wonderful insulation against the cool temperatures of the open ocean.

Sea turtle babies are, of course, not the only animals that seek refuge in drifting sargassum. In contrast to the barren open water, these expanses of seaweed—which often cover huge areas—are teeming with life. These free-floating sargassum habitats are the home of many small fish and crabs and offer the hungry young sea turtles a hearty buffet. And so the turtles grow quickly in relative safety, first expanding widthwise and then lengthwise. This growth pattern is a clever natural mechanism that ensures the little ones get too wide for the jaws of most predators as quickly as possible. In biological terms, this is known as morphological defense.

We still don't know, however, just how quickly small turtles grow in the first years of their lives or how many years in total they live in the open ocean. Estimates range from one to ten years and differ from species to species. What we do know is that leatherbacks and olive ridleys spend the rest of their lives—their juvenile life stages and their adult life stage—out in the open ocean. All other species of sea turtles move to coastal feeding areas in their older juvenile stage, which is when they are about the size of a dinner plate. Once their shells are 20 to 30 centimeters (8–12 inches) long, their size makes them more difficult to prey on and they are less likely to be swallowed by all the predators out there. They have obtained protection by virtue of their size, a so-called size refuge.

Juvenile turtles use the more abundant resources in their new feeding grounds to grow even more quickly. On average, most species increase in length by 3 to 5 centimeters (1–2 inches) a year, but right at the beginning of their time close to the coast and in especially nutrient-rich waters, they can grow up to 20 centimeters (8 inches) a year. Their chance of survival increases with every extra centimeter, and so the most important thing a young turtle can do is grow. Studies have shown that they often spend several years in especially productive feeding grounds before they move on.

These coastal nurseries are often found in more temperate zones than the turtles' tropical nesting beaches. Although the small turtles find plenty of food in higher latitudes, they encounter problems of a completely different nature. Sea turtles are "cold-blooded" or, in more scientific terms, ectothermic. That means they cannot generate heat internally, so their body temperature completely depends on the temperature of their surroundings. This is why sea turtles live mainly in tropical and subtropical zones, where the water is consistently above 20°C (68°F). Sea turtles have difficulty keeping warm when the water temperature drops below 10°C (50°F), and when the temperature drops rapidly, they can get cold stunned and die of hypothermia. A cold-stunned turtle first becomes weak and lethargic, then loses the ability to swim, and finally floats helplessly on the surface.

In shallow lagoons along the coast of Texas—which are important kindergartens for green turtles—increasing numbers of unexpected cold snaps in recent years have led to hundreds, even thousands, of cold-stunned young turtles. In a single mass event in 2021, more than twelve thousand (!) turtles were overcome by the cold and had to be gradually warmed up and given medical attention in various recovery centers. Not all of them survived. Up until then, many Texans had no idea there were any sea turtles along their coast, let alone that the turtles swam there in such numbers. As a result of global climate change, we must expect more frequent, and more extreme, weather events in the future, which likely will lead to more episodes like this.

Water temperature also affects the turtles' diet. Species such as the green turtle, which feeds mainly on macroalgae and plants such as seagrass as an adult, require the help of bacteria in their gut to break down the hard-to-digest plant material through microbial fermentation. These bacteria, however, must first be gradually acquired from the environment, and it seems they are most active in warmer temperatures. A new study that compares the food green turtles eat in different parts of the world establishes a clear link between colder water and more animal matter in their diet, even among adults. The inactivity—or potential lack—of these gut bacteria in colder waters likely explains why even species of turtles that are primarily vegetarian eat more animal protein when they are juveniles in the open ocean and for a short while after they arrive in their often still temperate coastal kindergartens.

Eating animal protein also helps speed up sea turtles' growth rate. Changing nutritional demands as the turtles mature are probably the reason young turtles change their feeding grounds every once in a while. Once they reach sexual maturity, adult turtles usually remain faithful to one feeding ground for the rest of their lives and return to it after migrating to the areas where they mate and lay eggs. Interestingly, sea turtles never stop growing, although the rate at which they grow drastically decreases after they reach sexual maturity. After that, depending on the species and the population, adults continue to grow in length from a few millimeters to a few centimeters (⅛ to 1¼ inches) per year.

AFTER THOSE OH-SO-SPECIAL WEEKS in Egypt, my first encounter with a sea turtle in the wild, and my restored courage to make my childhood dreams come true no matter what, I encounter an obstacle the moment I arrive back in Germany. Upon my return, I discover that the University of Hamburg is no longer accepting transfer students with my type of degree for their graduate program. I am studying at a time when all the universities in Germany are gradually switching to the system of bachelor's and master's degrees used in much of the English-speaking world. I belong to the last cohort of students receiving a German diploma under the old system. After a brief nervous breakdown and discussions with a few of my acquaintances and mentors from my time working at the dolphinarium, I decide to find a university that specializes in behavioral biology and animal ecology. At the end of the day, it should be up to me where exactly I will apply the basic concepts I will learn—and why not in the area of marine biology? And so, a few months later, I move to Würzburg, a decision that, combined with a happy accident, will determine the course of my life.

On the first day at my new university, I have an appointment with a professor and am waiting outside his office for him to come back from lunch. I notice a large black bulletin board almost falling off the wall under the weight of thousands of pieces of paper. To pass the time, I begin to read the notices. People are looking for roommates and selling books, but there are also postings for jobs and volunteer opportunities. Right at the bottom, a green logo with a tropical tree leaps out at me. I read the words "Costa Rica" and "leatherback turtles." What on earth is a leatherback turtle? Curious, I take a closer look at the notice. The German organization Tropica Verde e.V. is looking for research assistants for the sea turtle project run by its partner organization ANAI in Costa Rica, which is dedicated to saving the endangered leatherback turtle from extinction. Room and board are covered. Interested parties would have to commit for four months.

All of a sudden I have butterflies in my stomach. This sounds super exciting. It's exactly what I've been looking for. Then I see the date. The notice is from last year and the start date was in February. Right now, it's April. A feeling of deep disappointment courses through me. Even so, I write down the email address and send a note that evening asking if research assistants will be needed next year as well. The answer arrives a couple of days later: "Yes, we need assistants every nesting season." The butterflies in my stomach are back and begin to flutter wildly when I send in my application. A couple of months later, it's a done deal: I will travel to Costa Rica for four months to study sea turtles.

LEATHERBACK TURTLE

Dermochelys coriacea

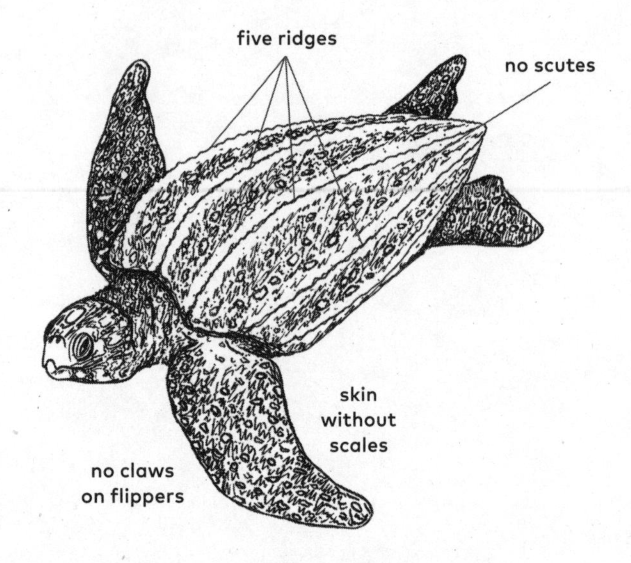

IUCN STATUS: VU—Vulnerable

CARAPACE LENGTH: 120–175 cm (47–69 in.)

WEIGHT: 300–600 kg (660–1,320 lb.)

MAIN DIET: Jellyfish

AGE AT MATURITY: 15–25 years

REMIGRATION INTERVAL: 2–3 years

INTER-NESTING INTERVAL: 10 days

NUMBER OF CLUTCHES PER NESTING SEASON: 6–9

NUMBER OF EGGS PER CLUTCH:
80 (60 in the Eastern Tropical Pacific)

INCUBATION TIME: 55–65 days

GOOD TO KNOW: Only species with a soft shell and no scales on its skin; deepest diver of all sea turtles; range extends into cold waters in northern and southern latitudes

– 3 –

COMING OF AGE

THANKS TO ITS RICH SPECIES DIVERSITY, Costa Rica is a hot spot for nature lovers and biologists, including marine biologists. Costa Rica has a landmass of barely 51,060 square kilometers (19,714 square miles), and yet, according to the Holdridge Life Zones map, this tiny country encompasses a whopping sixteen climate zones that range from hot and humid to cold and frosty and are home to a total of 5 percent of the world's biodiversity. Its tropical landscapes range from hot volcanoes to cool cloud forests, from dense rainforests to sparsely treed dry tropical forests, from emerald-green waves lapping at black-sand beaches bordered by coconut palms to remote waterfalls and rivers. Costa Rica is a veritable natural paradise with two coasts: the land gives way to the Caribbean to the east and the Pacific rolls up the beaches to the west. The waters on both sides are brimming with an incredible multitude of ocean creatures both large and small. Humpbacks come here to mate and give birth to their calves; big fish such as mahi-mahi and sharks swim by looking for well-fed prey. Many people living in the coastal villages practice traditional forms of fishing and depend on these local ecosystems, which unfortunately are not as abundant as they once were.

Along with the export of bananas, pineapples, sugar, coffee, biomedical equipment, and computer chips, tourism is one of the country's top sources of income. Costa Rica is celebrated around the world as an example of sustainable tourism and as a model for conservation. Although there are enormous resorts in a few regions, where most of the almost three million tourists who visit each year stay, when you travel around the rest of the country you have the place to yourself and can observe animals and plants at your leisure. You quickly acclimate to muddy jungle paths and deserted beaches and settle into enjoying the never-ending concert put on by howler monkeys, toucans, frogs, and armies of insects. You will have years' worth of stories to tell about encounters with the many different animals that keep the cacophony of sound going deep into the night—and might even find their way into your bed.

Five of the seven living species of sea turtles use Costa Rica's beaches and surrounding waters as places to find food, mate, and lay their eggs. An official decree that forbids the capture of female green turtles on beaches was introduced in 1963, and since then, Costa Rica has signed the Convention on International Trade in Endangered Species of Wild Fauna and Flora, commonly referred to as CITES. In 2002, the official decree was replaced by a law that specifically protects all species of sea turtles. The law states that using or eating sea turtle parts and products—whether eggs, meat, or other body parts—is punishable by one to three years in prison (with the exception of a legal egg-harvesting program, which I will discuss later). Unfortunately, this law has been enforced only sporadically, and poaching sea turtles and their eggs is still common practice in many parts of the country. This is particularly worrisome as all seven species are on the IUCN's Red List of Threatened Species. Therefore, many projects supported by thousands of volunteers from around the world have popped up in the last twenty years to protect and research these charismatic creatures.

ANAI, an organization based in Costa Rica, manages just such a project, and in 2007, I am on my way to join it. The project is based on the Caribbean coast in the tiny village of Gandoca. This is only the second time I have traveled more or less alone to such a faraway country, and the first time I have done so without any support or social safety net. When I spent a year as a foreign student in the U.S., there was an organization I could turn to and, of course, the host family I was staying with. Here in Costa Rica I have no one to rely on but myself. I don't know the organization that manages the project or any of the locals. For me, it is a journey into the unknown, and the whole trip feels a bit like a rite of passage into adulthood.

After a fourteen-hour flight and a sleepless night, I am sitting by the open window of a Costa Rican cross-country bus, overtired but also bubbling over with excitement. A taxi picked me up from the hostel early this morning and drove me to the bus station. There I met my English colleague Rachel and one of the locals who works for ANAI, José. Now I am being driven through a landscape that could have sprung directly from *Jurassic Park*. Fronds from enormous ferns extend from the roadside; small waterfalls break the dense greenery; trees covered in moss, bromeliads, and orchids form a tunnel of vegetation through which the bus is now racing at breakneck speed. I'm terrified at first, but after a while I follow the lead of my seemingly relaxed co-travelers and settle back in my window seat to enjoy the view. I can hardly contain my excitement. My French seatmate is now having to put up with my enthusiasm as I keep pointing out all the incredible things I'm seeing outside. He kindly plays along.

We wind our way higher and higher into the mountains, ending up in the clouds. We are immediately enveloped in mist. It's noticeably cooler and the trees are steadily dripping. Despite the steep slopes and hairpin bends, the bus does not slow down. Gradually, the hypnotic greenery outside my window works its magic and exhaustion catches up with me. It really isn't a good idea to fall asleep on a public bus in a foreign country with everything I own in my backpack or on my person, I think, as my eyes finally close.

Luckily, my seatmate is an honest man and none of the many professional backpack thieves are on my bus. I awake many hours later, covered in sweat but still in possession of all my belongings, when the bus erupts into chaos. We've just stopped in the village of Cahuita and half the passengers are getting off. The climate here is completely different. It's now hot and humid. I turn to look at José, who is sitting at the back of the bus with Rachel, and he signals that we still have about two hours to go.

Half an hour later we stop in the very touristy town of Puerto Viejo, and most of the remaining passengers get off before the bus continues toward the remote town of Sixaola. So far the road has had a meager covering of asphalt generously pocked with huge potholes. Now the asphalt gives way to gravel. The bus slows down a bit, but to make up for its reduced speed, it is now following a daredevil slalom course around enormous puddles that are lying in wait to coat the vehicle in red mud.

The views from my window still show impenetrable green on both sides of the road, but the vegetation has visibly changed since we descended from the mountains. The trees are larger and more diverse, their trunks spreading out into a checkerboard of roots, while in places a network of lianas and other aerial plants hangs down to the ground. I can see power and telephone cables paralleling the road strung between what look like extremely rotten wooden poles. As we drive by, I think I see an animal hanging from one of the cables. Was that a sloth? Unfortunately, my seatmate got off in Puerto Viejo and apart from my traveling companions, who both appear to be fast asleep, the only other passengers on the fifty-six-person bus are a woman and her young daughter up at the front. There's no one I can ask who might confirm my guess. I go back to staring out of the window.

We drive around a bend into a valley, and the lush tropical vegetation suddenly gives way to a sea of banana palms that appears to stretch endlessly to the horizon. When I look more closely, I make out at least one border to the right, where an enormous river snakes along the western perimeter of the plantation. Beyond that, the Talamanca mountain range rears up in the distance. For the next hour I see nothing but a monotonous monoculture of bananas and the occasional wooden hut on stilts.

After what seems like an eternity, we arrive at a row of trees on the left extending at a right angle from the main road. A sign with an arrow informs us that Gandoca lies 10 kilometers (6 miles) in that direction. José whistles loudly and the bus stops abruptly in the middle of the road. This apparently is where we get off. I pull my small backpack out from under the seat, climb down off the bus, and look around. Apart from banana palms and a somewhat dilapidated building, there's nothing to see for miles. We take our large backpacks out from the luggage compartment and have barely put them on when the bus roars off, opening up the view of the other side of the road and a waiting minibus. Once upon a time, it had probably been black, but it now has so many huge scratches and dents that have been roughly painted over that it looks a bit like a sickly, gray-black zebra. Next to the minibus a Costa Rican woman with wavy, chin-length hair is brandishing a sign with the name of the organization, ANAI. She bursts out into peals of laughter. Clearly José knows her, and the sign is completely unnecessary.

After all our things are loaded into the bus, we start on the last part of my twenty-four-hour journey to the isolated village of Gandoca. The gravel track through the banana plantations deteriorates until it has more potholes than drivable surface. Carmen, our driver, laughs out loud when she sees our faces and shouts in broken English over the rattling of the minibus: "A Costa Rican massage!" Rachel and I manage tight smiles. We're feeling a little queasy, especially when we arrive at the first of three river crossings and see that the "bridge" is nothing more than a couple of planks across the water. I can't help closing my eyes, opening them again only when we have made it safely across to the other side.

A few miles later, the lush growth of the rainforest replaces the dreary monotony of the banana plantations, and after another thirty minutes we finally arrive. Small dwellings fashioned from planks pop up on either side of the track. Most are on stilts or at least a raised foundation—because of the frequent floods here, we are told. There aren't many houses and no shops or restaurants, just two bars with signs advertising Imperial, the Costa Rican beer. We drive to the end of the road right on the ocean. There we stop and are offloaded along with our luggage, because the only way to complete our journey is on foot. The beach is bordered on the inland side by rainforest and we walk along a narrow track parallel to the ocean until, after a couple of hundred yards, we see a dark brown sign with yellow lettering announcing the ANAI sea turtle project. Next to the sign, a second path leads off to the left into the forest. We are almost there.

At last, we enter a clearing with two wooden buildings. One has two stories and is painted pale green; the other stands on six-foot stilts and was once yellow. The paint is peeling on both, and termites have eaten holes in the planks. There is no glass in the small windows, only mosquito netting, and several doors indicate about seven rooms in which I spy bunk beds. Behind the dormitories I see three smaller buildings, which I later discover are the toilets, a shower block, and a cookhouse with a screened porch.

This is palatial, I think. I had been expecting a backcountry camp complete with outhouse. Real beds with something resembling mattresses, running water, and a roof over my head? This is the height of luxury. I turn to Rachel excitedly. She is white as a sheet and stammers: "I'm not going to last a day here." It just goes to show how different our expectations were.

Several people are already gathered in the screened porch of the cookhouse to welcome us. Team members for two different projects are now all present and accounted for, including five international and five local assistants, a handful of volunteers, two biologists, and one student who is going to gather data for her master's thesis during this year's nesting season. In addition, there are three representatives of ANAI, who will lead our training sessions over the next few days. Two of my new colleagues accompany me to the room that is to be our home for the next few months: Julia from Canada and Marie from Ireland. We hit it off right away. I stash my backpack in a corner and set to work hanging my mosquito net. Following the recommendation in my guidebook to Costa Rica, I carefully tuck the edges under my mattress to prevent uninvited guests such as scorpions and other critters from crawling into my bed. After our evening meal, we all go right to bed because we have to get up early the next morning.

I spend a restless night. I lie in the upper bunk desperately trying to fall asleep. I am mentally and physically exhausted, but my mind is racing. Too many new experiences and not enough sleep over the past few days have put my brain into overdrive, and I feel drunk. I expected some feelings of euphoria now that I have fulfilled my childhood dream of traveling to a far-off land to research exotic animals, but none are coming. Everything feels so unfamiliar. The musty smell of wood all around me, the thin foam mattress under me, the night chorus from the rainforest, the soft breathing of my roommates. For the first time in my life, I am overcome with something that feels like homesickness. I toss and turn. Every time I turn over, the upper bunk shakes precariously, and the rotting wood creaks and groans.

Suddenly the door is yanked open and a dark figure enters the room. I sit up and manage to stifle a scream when I recognize Luis, one of the local assistants whom I met briefly earlier in the day. He is carrying a hammer. Without saying a word, he grabs one of the bedposts, nails it to the wall, goes to a second bedpost, and nails that to the wall as well. Then he disappears as wordlessly as he arrived, leaving me speechless. Apparently, my tossing and turning has made it impossible for him to sleep in the next room. I sympathize. His solution works beautifully and the bed is now rock solid. Even so, sleep eludes me. But I know that eventually this night, too, will pass.

With the first rays of the sun at 5:20 a.m., the howler monkeys in the *manzana de agua* tree outside our window begin to stir. If I hadn't encountered them the previous day, their ghostly screams—which can be heard from 16 kilometers (10 miles) away—would likely have made my blood run cold. This morning, however, they are a welcome Costa Rican wake-up call after an unsettled night. We get up and ready ourselves for our first orientation walk along the beach.

This is when I make my first big mistake. It's February and I have arrived directly from winter in Germany. Although I usually tan quickly, right now I'm still quite pasty. Because it's so early in the morning, it doesn't occur to me to apply sunscreen—something I will bitterly regret for some time. It will take bottles of aloe vera and cocoa butter before I finally get the multiple sunburned areas under control. Right off the bat, I acquire a healthy respect for the strength of the sun's rays close to the equator. To this day, sunscreen and a bottle of water are part of my standard equipment, whether I'm off on an hours-long hike or just popping round to the corner store. You never know what the day will bring—which is the second important rule out here. On this, my first morning, however, all I can think of as we set off for the beach are the things I will discover and how happy I am to finally be seeing more of my surroundings.

Everyone gathers at the entrance to the beach, some of us still looking extremely sleepy, and we all set off together. The sand is black and big waves are crashing onto the beach. I'd always imagined the Caribbean with turquoise water, white sand, and coconut palms. These palms do grow here, and even though the rest of my Caribbean fantasy doesn't come close to reality, I am impressed by the wild beauty all around me. We're walking north and Carolina, one of the biologists, is explaining that the beach stretches for almost 10 kilometers (6 miles). Every 50 meters (55 yards), there's a number written on a tree or a post intended to make gathering data easier. The numbers start at the north and continue to marker 160 in the south. This morning, we start at marker 57 by the camp and count off the numbers in descending order. The reason for this exercise is twofold. First, it familiarizes the newcomers among us with the beach by daylight before we have to walk it in the pitch dark looking for sea turtles. Second, it's an opportunity to check whether all the markers from the previous season are still there. We have brought along a long measuring tape, two buckets—one of white paint and one of black—and reflective tape so that we can measure off distances and replace markers right away should any be missing. There are more of us than are needed to get the work done, so little clusters of people form at the edges of the group, talking animatedly and getting to know each other better as we move along the beach at a sloth's pace.

Our training continues over the next few days. I find it quite challenging because much of it is in Spanish, a language I don't know very well. Through English translations and the graphics in the PowerPoint presentations, I get the gist of what our instructors are telling us. We are going to work mainly with female nesting leatherback turtles and with smaller numbers of green and hawksbill females, all of which come to the beach under cover of darkness to lay their eggs, and we are going to do our best to protect the females and their eggs from poachers. We're also going to gather data on population size, the numbers of eggs laid, and the overall success of our conservation measures.

I don't know it yet, but the stage when females nest is the most intensively researched stage of sea turtles' lives. You might even say that in comparison with their male counterparts and turtles in all other stages of life, nesting females are over-researched. The reason, of course, is that sea turtles spend most of their lives in the ocean and that presents researchers with a number of challenges. It is costly and requires a great deal of effort to observe male and pre-adult sea turtles, and to gather data and samples from them. We have to follow them by boat first to the nurseries where the hatchlings get their start in life, which are far off the coast, and then to the sometimes difficult-to-reach coastal areas used by many species of sea turtles as kindergartens and feeding areas. Once we find turtles, we either have to get them out of the water using nets or other capture methods, or we have to go into the water with them to study them up close. All this is possible only with special gear, and boats, fuel, and equipment are extremely expensive. Nesting beaches, in contrast, are easy for us to access, and we can observe the turtles up close and on foot without too much effort on our part. That has led to a wealth of data about females over the past sixty years—their nesting preferences, nests, eggs, and babies—but also to a real dearth of data on juvenile and subadult turtles in general and on males in particular, which never leave the water once they have entered it after hatching. The big exception here is Hawaiian green turtles, also called *honu*, which are the only population of sea turtles known to climb sandbars or drag themselves onto beaches to bask in the warm rays of the sun.

As I have never seen a nesting sea turtle in my life, let alone a leatherback, I have a lot to learn. The lectures on the biology and ecology of sea turtles and the principles and reasoning behind our conservation efforts are broken up with plenty of hands-on activities. Using a cardboard box and a watermelon, we learn how to attach metal flipper tags to a sea turtle and insert a microchip. Out in the field, we are going to attach the external metal tags to either the front or hind flippers, depending on the species, and inject the microchip, which is about the size of a grain of rice, into the turtle's shoulder. We practice measuring sea turtles using sand sculptures on the beach.

It's also important for us to practice how to behave properly around sea turtles. We learn never to dance around directly in front of them but to always stay behind them in their blind spot. According to the official guidelines, we are to use only red light in the presence of sea turtles, because this light disturbs them the least. Many marine animals don't see red at all or only poorly because the long waves of red light from the sun are the first to be absorbed in water and don't penetrate deeper than 10 meters (33 feet) below the surface. With a few exceptions, the underwater world doesn't recognize the color red. It's also much more pleasant for us to switch between red light and darkness. It takes about half an hour for our eyes to switch from seeing in bright light to seeing in the dark. If we kept switching on white light, which is bright, our eyes would adjust to its wavelength, and when we then turned off the white light, we would end up stumbling about in the darkness not seeing anything at all. Red light, in contrast, does not interfere with our night vision.

Another crucial factor in the success of our conservation efforts is knowing how to dig and shape the artificial nests where we relocate the eggs we rescue. Different species of sea turtles dig nests of different shapes at different depths. The nests must be large enough to hold all the eggs in a single clutch. We practice digging nests for different species on the beach. This sandy task is accompanied by the psychological pressure to get things just right, because if a nest is too small or too large, too deep or too shallow, the eggs will die. Each of us must complete three acceptable nests for each of the three species of turtle that come to Gandoca to lay their eggs.

I have no problem getting the green and hawksbill nests right, but my leatherback turtle nests keep collapsing when I try to enlarge the egg chamber. Andrey, who up until now has been very reserved, watches me for a while and finally comes to my aid. He feels the inside of my nest and delivers his crushing verdict in English: "You started all wrong!" Then he shows me how he begins a nest so that it is stable from the outset, and how he removes sand by carefully scraping it out from the inside without having the whole construction fall in. I watch, impressed at how skillfully he moves his hands. He makes it look so easy. When it's my turn again, I dig according to his instructions and finally construct my first passable leatherback turtle nest.

Our training includes our first night patrols. Initially, we are led by experienced local assistants. On my very first night, I can barely see my hand in front of my face and I stumble blindly behind Luis—or rather behind the two white stripes of the tiny Costa Rican flag printed on his left sleeve. I keep running into logs. How on earth am I supposed to find a sea turtle in the pitch dark? I feel panic slowly rising inside me. But when the other international assistants and I share our experiences the next day, I am relieved when it quickly becomes clear that I am not the only one having problems with night blindness. Carolina, our biologist, reassures us: "Your eyes and your brain will get used to the dark in a few days. Just wait. You'll soon be able to see better at night." She's right.

On my second patrol, I'm already imagining that the darkness is not quite so impenetrable and that I can make out more and more of my surroundings—which is a great relief. But on the third night, Andrey, who is always reserved, concise, and to the point, informs me: "You're leading the patrol tonight." What? I'm not nearly ready! He's not interested in hearing excuses, however, and so the other two assistants and Andrey traipse along behind me in single file. I stop at every large dark object on the beach or in the water, trying to make out what it is. Every time, Andrey calls from the back, "That's not a turtle!" I can hear the amusement in his voice. What a joker. On our second break, he asks, "Why are you walking so slowly? You're stopping every five meters." I shrug and admit, "I'm scared to death of missing a turtle." He laughs out loud. I look at him, annoyed. What's so funny about that? Once Andrey has wiped the tears of laughter from his eyes, he replies in all seriousness, "You would have to be blind to miss a leatherback turtle." I suppose I should believe him, but I'm still a bit worried.

With the benefit of hindsight, I can confirm that Andrey was absolutely right. It is extremely difficult to miss a leatherback turtle. Even if you miss seeing the turtle itself, you can see its enormous tracks from a long way off even in the dark, and if you don't see the tracks you will feel them with your feet when you walk over them. However, in those early days I didn't have any experience with turtles and I was afraid of doing something wrong, with potentially disastrous consequences. Luckily, all this would change in the months to come.

A FEW WEEKS after successfully completing our training, I am leading patrols on my own. By now I have seen several leatherback turtles nesting, and under supervision I've practiced on real live turtles, learning how to attach metal tags to their hind flippers and how to inject a microchip. I am now walking the beach six nights a week for five hours at a time to find nesting females and make sure their eggs are safe. During the day, I am busy organizing materials for the two hatcheries on the beach and training new volunteers. In the afternoons, I join the rest of the team to clear plastic and driftwood from stretches of the beach. I have one free day and one free night a week. Gradually, I slip into a routine and the days begin to merge. I find it most relaxing to abandon myself to the flow of the project without having to think too much. We are fed three meals a day and lack for nothing—except chocolate. I experience my nightly wanderings under stars twinkling in the sky while listening to waves crashing on the beach as a form of meditation. It's been ages since I have had so much time to myself and time to contemplate God and the world.

One theme that especially occupies my thoughts in these first four months in Costa Rica is the clear differences between how boys and girls—and men and women—behave and are treated. Although I am here to study sea turtles, I cannot ignore this phenomenon. It's everywhere and affects both me and my work. In Gandoca, as in other rural areas of Costa Rica, gender roles are still very traditional. The woman stays home with the children, cooks and cleans, and keeps the family together, while the man is the breadwinner—though in this country he is earning money not for bread but for rice and beans. These antiquated attitudes are reflected in our sea turtle project as well. On the local side, men dominate from management down to the assistants. There is one female Costa Rican assistant and the head biologist, Carolina, who is Chilean but holds a Costa Rican passport. Apart from the cook and the cleaning crews, there are no other Costa Rican women in the camp.

Even though it's becoming more common in Costa Rica for young people to finish high school even in rural areas, few go on to college. Instead they enter the workforce right away, which here in the Caribbean often means working on the banana plantations. Even fewer women than men go on to college, as women's education is often not regarded as important. Most of them do finish high school, however, thanks to an elaborate system of grants, where the government supports all parents of school-age children by paying them a monthly stipend. And yet too often, I overhear conversations that go like this: "You don't need to be clever, just pretty, so you can find a good man who will look after you." Not good advice if women are to be financially independent and self-sufficient, and able to make their own decisions about whom to spend their lives with and what those lives will look like. In addition, young men often exploit the women's need for love and security and the instilled subordination toward men to pressure them into having sex far too early, which naturally leads to teenage pregnancies. All this draws young women into a vicious cycle from which they cannot escape. With children to care for and lacking education, they are then completely financially dependent on their partners, no matter how those partners treat them.

This is in stark contrast to my own upbringing and the values instilled into me by my parents, by German society, and especially by my two emancipated aunts. Of course, there is sexism in Germany. But the misogynistic attitudes and behaviors there are often more subtle, and the younger generation, in particular, is becoming more conscious of them. When I was a child, I always thought adults were not really listening to me because I was still a child. As I got older, I thought it must have something to do with me personally. Perhaps I was just not expressing myself clearly enough. It was probably one of the most painful moments of my adolescence when I realized that some of these reactions were in all likelihood simply because of my sex.

As I've already mentioned, I was a wild child. I climbed trees and tore my clothes. I was brave and good at sports. I dove into the water from the high diving board and from cliffs. In sports, I was proud that I was one of the first to be picked for teams and that I could run faster than many of the boys. Sometime during puberty, however, all that changed. The boys no longer wanted to play with me, and I wasn't girly enough for the girls. I was called a "tomboy" on more than one occasion. These changes were confusing and triggered some insecurities in me. For some reason I didn't understand, I belonged to neither side and was treated by society like a girl who didn't fit the norm.

I can thank a teacher of mine for one of the aha moments when I realized this had nothing to do with me personally. Whenever a girl put up her hand and gave the correct answer, the teacher would acknowledge her with a nod. If a boy then put up his hand and gave exactly the same answer, the teacher would reward him with high praise. The injustice infuriated me. And it made me wonder what that said about how society as a whole valued the opinions of girls. This realization led to deep doubts about how I should live my life as a girl and what that might mean for my dream.

It didn't help that my father loved to tell blonde jokes, as many people did back then. I was quite young when I caught on to the fact that people—make that men—believed that blonde women were stupid. I can't express in words how much that affected and still affects how I think and behave even as an adult. My biggest worry at university was always that I might be labeled a "dumb blonde" and not be taken seriously. That worry about appearance influenced many of my decisions right down to my wardrobe choices. For my undergraduate examination, for example, I arrived wearing a turtleneck and glasses, which I detested. I made sure I was extra well prepared for every lesson and every lecture and had an answer ready for every single question.

In Costa Rica, my experiences with sexism and the cultural machismo were even more extreme. In the very first week, I had to listen to comments about my not being cut out for this work. Apparently some of my Costa Rican colleagues had laid bets on how soon I would pack my bags and leave. Over the years, male bosses also kept letting me know—at times directly, at other times less so—that I should present and express myself less assertively because men don't like women telling them what to do. It would be better if I were to persuade them that whatever had to be done had been their idea from the beginning. For the first few years, I tried to adapt and keep my personality in check. That didn't work well. Finally, I became ill and almost lost myself to self-doubt. That was the moment I decided I needed to take charge and stop depending on my mainly male bosses.

Sometimes I wonder if growing up is the same emotional roller coaster for sea turtles as it is for people. Probably not—perhaps because they usually have much more time than people do before they become sexually mature. Depending on the species, they reach sexual maturity sometime between the ages of fifteen and forty-five. The journey they must take to reach this milestone, however, is long and arduous, and a lot can go wrong.

Sea turtles—after extensive travels and many growth spurts—reach a certain size when they, like people undergoing puberty, experience some bodily changes. However, aside from the length of their carapace, the external signs that turtles are becoming sexually mature can be seen only in males. The male's tail elongates, the claws on his flippers become longer and stronger, and the bony plate on his underside—the plastron—starts to become slightly concave. Apart from that, there are few external differences between males and females. Both are about the same size, and neither sex is more colorful or more eye-catching in any other way.

Before sea turtles reach puberty, it's impossible to tell males and females apart just by looking at them. If you want to know for certain whether a baby or juvenile turtle is male or female, you have to look internally. This means either killing the turtle and dissecting it, or performing a laparoscopy, where you insert two cameras into the body cavity so you can take a closer look at the internal sex organs to see if the turtle has testes or ovaries. As one method is gruesome and the other is highly invasive, researchers have spent years trying to find a way to analyze the concentration of hormones such as testosterone and anti-Müllerian hormone in the blood. The results are promising, and hopefully blood tests will soon become the standard way to sex sea turtles.

Although there's not much to see on the outside as sea turtles undergo puberty, things are changing on the inside. The testes of males begin to pump out higher levels of the steroid hormone testosterone, which stimulates the softening of the plastron and the formation of secondary sexual characteristics such as the elongated tail. It also stimulates the growth of the seminiferous tubules within the testes, where the sperm are formed in a process called spermatogenesis. In females, hormonal changes increase the size and change the structure of both the ovaries and the oviducts. In comparison with the ovaries of immature or pubescent females, those of mature females typically have enlarged stroma, the connective tissue that supports the organ. Each ovary is attached to a coiled oviduct with a diameter of at least 1.5 centimeters (0.6 inches), which hangs down into the body cavity next to the ovary. The oviducts of mature females are extremely long, measuring 4 to 6 meters (13–20 feet). Other signs of sexual maturity include the presence of ovarian follicles and the start of vitellogenesis, the process by which the yolk is formed and accumulates in the ovum.

The differences in size and weight between a baby turtle and an adult are striking: size and weight increase tenfold and one-thousandfold respectively. These enormous increases are critical to the babies' survival and are the reason scientists don't expect many babies to reach adulthood but have high expectations for the survival of older juveniles, subadults, and adults. By the time they are fully grown, sea turtles have few natural enemies other than sharks and jaguars. If only there weren't any humans. As I will explain in later chapters, with our lifestyle we are killing hundreds of thousands of sea turtles every year, especially turtles that have reached sexual maturity despite all the dangers. This is the reason the loss of every adult turtle is a tragedy for the species as a whole.

The females we are trying to protect on the beach are among the few that have made it to sexual maturity. Even though the future is uncertain, they are doggedly laying their eggs night after night, and I have the privilege of watching them as I learn more and more every day. On a night toward the end of my first visit to Costa Rica, I am walking the beach with two other volunteers, when we suddenly come upon not one but three leatherbacks practically climbing over one another as they race to get their eggs laid. What an incredible sight! I position each of my volunteers behind a turtle to catch the eggs because I'm going to have to move the eggs from at least two of the nests. Meanwhile, I have my hands full gathering data, marking the turtles, transferring eggs, and giving the volunteers a stream of instructions and explanations as they carry out their tasks. The next few hours fly by. By the time we leave the beach at the end of our shift at midnight, I am bathed in sweat and over the moon. On our way home, I see that the bar in the village is still open. My stomach is rumbling loudly, so I order a beer and the only edible item on the menu: a Snickers bar. In the years to come, I will order both whenever I celebrate a successful night.

GREEN TURTLE

Chelonia mydas

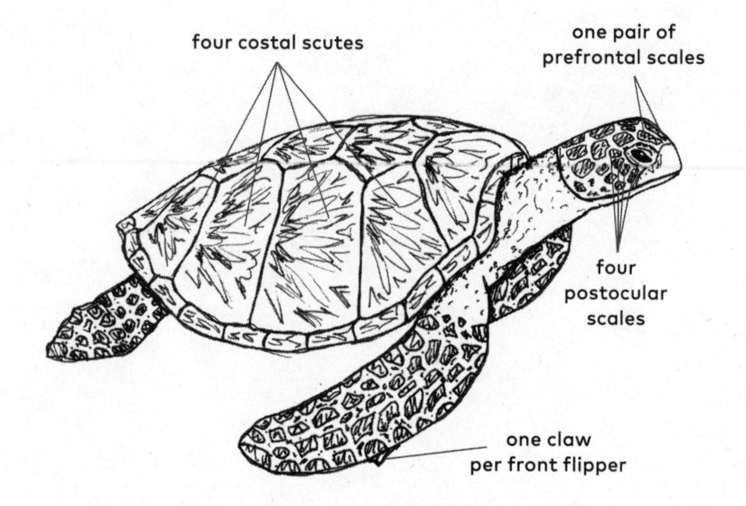

IUCN STATUS: EN—Endangered

CARAPACE LENGTH: 80–120 cm (31–47 in.)

WEIGHT: 120–220 kg (265–485 lb.)

MAIN DIET: Seagrass and seaweed (adults)

AGE AT MATURITY: 25–45 years

REMIGRATION INTERVAL: 2–5 years

INTER-NESTING INTERVAL: 12 days

NUMBER OF CLUTCHES PER NESTING SEASON: 3–5

NUMBER OF EGGS PER CLUTCH: 110

INCUBATION TIME: 50–60 days

GOOD TO KNOW: Almost exclusively herbivorous as adults; fat is greenish because of their mainly plant-based diet, hence the common name

– 4 –

YOU ARE WHAT YOU EAT

IF YOU'VE EVER BEEN lucky enough to meet a sea turtle face to face, you can probably remember the feeling of looking into the wrinkled but wise face of an ancient, toothless woman. The main reason sea turtles are associated with wisdom is probably because they look so old and because they live such long lives. I, however, like to think of them as wise because their evolutionary history stretches back eons.

Taxonomically, sea turtles belong to a superfamily (Chelonioidea) in the order of turtles (Testudines). Together with snakes, lizards, and crocodilians, turtles belong to an obsolete class called Reptilia. Obsolete because recent molecular analyses have made it clear that reptiles do not belong to a monophyletic group; that is, they don't have a common ancestor. For example, in evolutionary terms crocodilians—an order that includes crocodiles, caimans, and alligators—are more closely related to birds than they are to other orders of reptiles. This new research has led to reptiles being classified, along with birds and dinosaurs, in a group known as Sauropsida. However, even in scientific circles, we often talk of Reptilia when what we really mean is Sauropsida.

The sea turtle superfamily is divided into two families: the hard-shelled Cheloniidae and the soft-shelled Dermochelyidae. Only a single species of soft-shelled sea turtle still exists on this planet: the leatherback turtle, *Dermochelys coriacea*. All other living species belong to the family Cheloniidae. Over the years, I've been lucky enough to meet three members of the latter family in Costa Rica: the green turtle, *Chelonia mydas*; the hawksbill turtle, *Eretmochelys imbricata*; and the olive ridley turtle, *Lepidochelys olivacea*. The loggerhead turtle, *Caretta caretta*, I encountered years later in the U.S. for the first time. Although a handful of loggerheads nest every year on the most northerly Caribbean beaches of Costa Rica, their nests are usually found in more temperate latitudes.

The two other living species of sea turtles are endemic, which means their distribution is restricted to a limited geographic area. The Kemp's ridley turtle, *Lepidochelys kempii*, is mainly found in the Gulf of Mexico and nests in Mexico and Texas, which is where I eventually experienced one in the wild. It is the sister species to the olive ridley turtle. The flatback turtle, *Natator depressus*, can be found only in coastal waters north of Australia and south of Indonesia and Papua New Guinea, and so far it is the only species I have never seen live in my sixteen years researching sea turtles.

The evolutionary story of sea turtles reaches back to the Cretaceous. The first sea turtles split from the ancient turtle lineage more than 120 million years ago. To put this into context, most dinosaurs died off 66 million years ago during the Cretaceous-Paleogene extinction event. Fossils from the most ancient known sea turtle species were not discovered until 2015. They come from *Desmatochelys padillai*, a turtle that was more than 2 meters (6.5 feet) long and swam in the depths of the prehistoric ocean until it went extinct. One of the first and largest fossil finds of a prehistoric sea turtle belongs to *Archelon*, a giant that was more than 4.6 meters (15 feet) long and must have weighed more than 2 tonnes (2 tons). For a long time, we thought *Archelon* was a direct ancestor of modern leatherback turtles, but that has since been refuted.

There have likely been more than thirty different species of sea turtles over the millennia, of which only seven still exist today. The oldest and largest is "my" leatherback turtle. It was the first species I had the honor of observing while it laid its eggs, and for years it has been the focus of many of my projects. Both its English and its scientific genus name (*Dermochelys*: *derm*, Greek for "skin"; *chelon*, Greek for "turtle") reflect that it doesn't have a hard, bony carapace but a soft one made from cartilage covered with leathery skin. It seems to be descended from a different lineage than the other modern species of sea turtles. Scientists estimate that the leatherback turtle has existed in its current form for about 90 million years, which means the great-great-great-great-great-great-grandmothers of leatherbacks alive today probably knew a few dinosaurs. The family Cheloniidae split off from a common ancestor about 60 to 30 million years ago and evolved into the remaining six modern species. To this day, I am awestruck when I watch a sea turtle laying her eggs and think how lucky I am to be sharing the beach with such a prehistoric creature.

ONE REASON THE SEVEN EXTANT SPECIES have survived the last few million years is probably their diet, or more precisely their adaptability when it comes to food. The species that went extinct had developed overly specialized eating habits, which caused problems in a constantly changing world. In contrast, the modern sea turtles were far more flexible and could switch food sources when necessary. This allowed them to avoid food shortages and competition for food. The exception is the leatherback turtle, which likely never had much competition because of its specialized diet of jellyfish.

But why hasn't one of the seven species displaced the other six? Generally speaking, the ecologies of the different species are quite similar. They live primarily in subtropical and tropical waters and mostly feed close to the coast. And it is common for several species to share the same beach to lay their eggs. It seems, therefore, that they would be in direct competition for nesting and feeding areas. However, over the millions of years they have existed side by side, slight differences in their spatial occupation of the marine habitat and barely overlapping food sources have led to a successful coexistence and to the development of specific ecological niches. Diet is a huge part of the identity of the different species, and diversifying their diets has been the key to sea turtles' success.

For example, the leatherback is the only species of sea turtles that spends most of its time in the open ocean even after it is sexually mature because nutrient-rich currents offer perfect conditions for its main prey, jellyfish. Leatherbacks are among the few animals that help keep jellyfish populations in check. The latter are multiplying exponentially as our oceans are increasingly being polluted by excess use of fertilizers in agriculture, and jellyfish have become a real nuisance in tourist areas and for fisheries. The dietary preferences of leatherbacks are a natural way to control them.

Leatherbacks are also the only species of sea turtles able to dive to a depth of more than 1,000 meters (3,300 feet) to find food and avoid predators. All sea turtles are pretty good divers, of course. They fill their lungs with air when they want to float, and, depending on the species, they can spend from forty-five minutes to two hours underwater when they are resting on the ocean floor protected by cliffs or coral reefs. For routine activities when they need more oxygen, they dive for shorter intervals—from four to ten minutes—and in between they come to the surface for a few seconds to breathe. A turtle in distress, in contrast—for example, one that finds itself caught in an abandoned fishing net drifting in the ocean, a so-called ghost net—will quickly exhaust the supply of oxygen it has stored in its body and might well drown. Due to their hard shell, which cannot withstand the pressure in the ocean depths, Cheloniidae generally don't dive very deep. They usually hang out in the upper 100 meters (330 feet) of the ocean. Only the soft shell of leatherbacks can accommodate the increased pressure in the ocean depths with no problem. The deepest documented dives of leatherbacks are more than 1,200 meters (3,900 feet), and, interestingly, they are usually recorded near nesting beaches and not in their feeding areas. Scientists suspect that these deep dives are a way to evade sharks, as turtles mostly display this behavior in areas known for the presence of tiger sharks.

Leatherbacks' preferred feeding areas are in the cold parts of the ocean in the high northern latitudes and deep southern latitudes, where plankton and jellyfish thrive. That's another unusual thing because sea turtles, as ectotherms, are normally found in waters with temperatures close to their ideal body temperature of 25°C (77°F). Turtles are dedicated sun worshippers, bobbing up and down on the water surface and letting the sun shine on their shells for hours. The temperature of the waters in which leatherbacks hunt, however, is many degrees lower than the body temperature they need for a functioning metabolism.

Luckily, leatherback turtles are masters of thermoregulation (some scientists even consider them endothermic). Over the course of their evolution they have developed the ability to stabilize their body temperature at around 25°C even when swimming in waters colder than 10°C (50°F). They have a thick layer of fat, similar to the blubber of whales and dolphins, that functions like a winter jacket, protecting their bodies from the cold. Even their esophagus is wrapped in a layer of fat, which prevents the food they ingest from cooling their body from the inside. It must be said that this winter jacket of fat is not particularly practical when the turtles migrate to tropical areas to mate and lay eggs. They burn through a lot of their stored fat on the journey to their nesting beaches, but females still have an impressive layer of fat on them when the egg-laying season begins.

This is where another adaptation comes into play. Leatherback turtles can selectively open and close the blood vessels between their fat layer and their skin. In cold waters, these vessels are mostly closed so that the warm blood circulating through their system remains in the interior of their bodies and isn't cooled by the surrounding water. In warmer waters, however, the turtles open these blood vessels to release heat and avoid suffering heat stroke. There's just one exception: their flippers. Even though the flippers must plow their way through ice-cold water, they are not encased in a layer of fat and the turtles cannot—and must not—shut down the blood vessels in them. The blood must circulate so they can keep using their flippers to propel themselves through the water and so their limbs do not wither away and die.

Leatherback turtles can keep their flippers supplied with blood and their core temperature stable even in cold water thanks to what is known as countercurrent exchange, which is a mechanism describing the exchange of heat or chemicals between two liquids or gases. In the case of the turtles' flippers, it prevents heat loss and therefore conserves energy. This is how it works. On its way into the flippers, warmer arterial blood is cooled by cooler venous blood returning to the body so that by the time the arterial blood reaches the flippers, it is basically the same temperature as the water outside. As the cooler blood returns to the body via the veins, it absorbs heat from the arterial blood heading to the flippers so it's warmed up by the time it gets back into the turtle's body. This system (also found in penguin feet and in the paws of Arctic foxes, for example) ensures that the turtle doesn't need to expend much energy bringing the blood back up to its core body temperature. It does mean, however, that leatherback turtles always have cold hands—or should I say, flippers.

When it comes to retaining heat, the leatherbacks' enormous size also works to their advantage because they have a small body surface compared to their large body volume. That means that although a leatherback turtle is huge, a relatively small amount of its body is in contact with its environment, so it loses less heat. This phenomenon, also known as gigantothermy, means that, thanks to its relatively smaller surface area, a large, bulky ectotherm finds it easier to maintain a constant, adequately high body temperature.

Another recent groundbreaking discovery has revealed that leatherback turtles are even capable of generating a bit of heat on their own as a byproduct of digestion. This is highly unusual for a reptile, especially one that eats almost nothing but jellyfish, which in turn consist of almost nothing but water. Just how the turtles can maintain their enormous bulk on such a watery diet is a secret they haven't yet shared with us.

In contrast to leatherback turtles, hard-shelled sea turtles in the family Cheloniidae spend most of their adult lives in coastal tropical and subtropical waters above the continental shelf, although some populations of olive ridley turtles can be found in deeper waters, even after they have grown up. Olive ridleys are true omnivores, which puts them at the opposite end of the spectrum of dietary specialization from leatherbacks. From algae to fish, olive ridleys consume everything that floats or swims their way, and no one food source is more important than another. Their sister species, the Kemp's ridley, prefers to stay near the coast, where the adults feed almost exclusively on blue crabs. A few reports have documented them also occasionally feeding on jellyfish and tunicates, which are marine invertebrates somewhat like jellyfish.

Green turtles are omnivores only during their youth. Greens are the only species of sea turtles that mainly eat plants as adults. In only a few populations do the adults continue to predominantly eat animal protein. In most populations, adults mostly feed on seagrasses and algae. Every now and then, they will feed on jellyfish and other soft-bodied marine invertebrates, but these make up a relatively small part of their diet. All this plant material turns their body fat slightly green, hence their common name.

Green turtles with their vegetarian diet play an important role in maintaining seagrass meadows. They are highly selective grazers, and as they feed, they create natural disturbances just as storm waves do. These disturbances ensure continuous regrowth of the seagrass while preventing some of the faster-growing species from taking over. The biochemical composition of younger leaves of seagrass makes them easier to digest. Younger leaves also contain more nitrogen, which is particularly important for herbivores as this nutrient helps them synthesize protein. Older leaves contain more structural material, which makes them harder to digest. Through selective grazing, green turtles ensure seagrass meadows are full of young leaves that are investing more of their available energy in nitrogen and carbohydrates and less in structural material. This benefits not only the green turtles but also a host of other marine herbivores, large and small. In areas where the populations of green turtles have declined and seagrass growth has not been kept in check by grazing, the blades of seagrass are older and larger, and over time a thick layer of dead leaves accumulates in the seagrass beds.

Changes in seagrass meadows cause problems for the turtles as well. Marine traffic has transplanted the seagrass *Halophila stipulacea* from its original locations in the Indian Ocean and the Red Sea to the Caribbean. This invasive species has been displacing native species in Caribbean seagrass meadows for years. Green turtles can eat *Halophila*, but its nutritional value is low. Up until now, few Caribbean sea turtles have been eating this new-to-them seagrass. Studies show, however, that it is likely to be more easily digestible and this could balance out its lower nutritional value because it would take less energy to break it down. Right now it looks as though *Halophila* could become the dominant species in the Caribbean unless people step up to actively control its spread. If green turtles can switch to feeding on this new species without adversely affecting their health, its dominance might not be a problem for them. But only time will tell.

The hawksbill turtle has chosen the opposite tactic. It's a carnivore that has specialized in feeding on sponges. In some areas, like the Caribbean, sponges make up as much as 95 percent of the turtles' diet; in others, it is less but sponges are still their most important source of food. Sponges can threaten reef-building corals by growing faster than corals and eventually replacing them. Thanks to their dietary preferences, hawksbill turtles are true coral gardeners and help protect coral reefs, which are also essential for the survival of many other critically important organisms.

The loggerhead turtle is also a carnivore, but it feeds on a wider range of prey and doesn't specialize in a single group of animals. Its name in both English and Spanish (loggerhead and *cabezona*) aptly describes its enormous head and super-strong jaws, both indicators of its dietary preferences: invertebrates with hard shells, which it can easily crack open. Adult loggerheads feed mainly on crabs and a wide variety of other marine animals such as mussels, sea snails, and a class of free-floating tunicates called thaliaceans. The turtles' jaws, with a bite force of more than 2,100 newtons (472 pounds-force), pose a danger not only to hard-shelled invertebrates, but also to sea turtle researchers. Some of my colleagues have lost parts of their fingers to loggerheads when they came too close to their beaks in a moment of inattention.

We don't have much information about the dietary preferences of Australian flatback turtles. They are unique among sea turtles in that their young do not swim out into the open ocean. These turtles spend their entire lives above the Australian continental shelf and close to the southernmost islands of Indonesia and Papua New Guinea. The small amount of available information suggests, however, that young flatbacks, like the juvenile stages of other species, mostly feed on zooplankton in the upper part of the water column. Adults switch to foraging on the ocean floor, hunting down soft-bodied, slow-moving marine invertebrates such as soft corals, sea pens, sea cucumbers, and jellyfish.

In the past, researching the eating habits of sea turtles was a rather nasty business. To draw informed conclusions about the types of food they ate and how much of each they fed on, you had to inspect their stomach contents. In the past century, scientists who pursued this line of inquiry were, therefore, experts in gastric lavage and necropsies—which is a fancy way of saying forcing live turtles to vomit so you could poke around in whatever came up, and slicing open dead ones. Not for those with sensitive stomachs (no pun intended!). This research did, however, yield a great deal of useful information and considerably advanced our knowledge of sea turtle ecology.

Despite these fruitful means to collect data, I'm happy to report that these days we have more elegant and less nauseating methods at our disposal. For example, for my PhD research, I examined stable isotopes in skin samples to answer questions about sea turtle eating habits. These analyses can be done anywhere, and just a small sample of tissue allows me to travel back into a sea turtle's past. Depending on whether I take samples of blood, skin, or keratin of the carapace, isotopes of carbon, nitrogen, oxygen, and other chemical elements within the tissue allow me to trace an individual turtle's food intake back for several weeks or even years and draw conclusions about not only its dietary preferences but also its preferred feeding grounds. For example, the ratio of the two carbon isotopes $^{12}C/^{13}C$ in a tissue sample tells me how close to the coast or how far north or south of the equator the sea turtle has fed. The ratio of the nitrogen isotopes $^{14}N/^{15}N$ provides information about the individual's position in the food chain and whether, for example, it is a herbivore or a carnivore. Isotope analysis also bypasses the limitations of stomach-content studies, where you usually have to go out and sample the turtles in their feeding grounds.

Another fascinating topic is how exactly sea turtles detect their food underwater and eat it. Over short distances in good visibility, potential food can be located by sight. Over longer distances in murky water, it helps to have a sharp sense of smell to sniff out prey and follow its scent trail. When you look at a sea turtle's brain, you notice that it's fairly small for an animal of its size, but the area devoted to olfactory stimuli—a.k.a. smells—is relatively large compared with all the other areas. Consequently, sea turtle brains suggest they might not be particularly intelligent or capable of extensive learning, but they are equipped with a sensitive nose.

Over millions of years of evolution, sea turtles have successfully mastered the art of eating underwater without swallowing mouthfuls of sea water. The esophagus of all species is coated with keratinized spines oriented toward the stomach. When a sea turtle eats, it ingests sea water along with its food. The food gets impaled on the spines and stays put as the turtle ejects the sea water (mostly through its nose). The turtle can then swallow its meal and move it on down into its stomach without a side serving of salty water. This clever adaptation has unfortunately proved to be a problem for sea turtles in recent years. In the modern age, what was once an evolutionary advantage has now become a disadvantage: the mechanism that makes it possible for them to feed efficiently underwater prevents them from spitting out ingested plastic because it gets stuck in their spines. For many years now, plastic in the stomachs of sea turtles has been an unhappy part of their everyday lives and of the everyday lives of sea turtle researchers, who have been finding plastic bags in the stomachs of leatherback turtles since the 1970s and 1980s.

PEOPLE HAVE ALWAYS HAD a significant impact on the lives of sea turtles, and our relationship with them can be described as ambivalent at best. On the one hand, sea turtles are often revered in our stories and legends. For example, in Hindu mythology, Earth is carried by four elephants standing on the back of a turtle, which is sometimes depicted as a sea turtle and sometimes as a tortoise.

In China, the goddess of creation, Nüwa, cut the flippers from a sea turtle and used them to hold up the sky. This is based on the ancient Chinese concept that the flattened plastron and convex carapace of a turtle represent the flat earth and the domed sky. In Polynesian culture, the sea turtle, or *honu*, represents good health, fertility, longevity, stability, peace, and tranquility. A Hawaiian legend tells of Kauila, a huge sea turtle goddess, who transformed herself into a human girl to play with the *keiki*, the children, and protect them. In the culture of the Lenape and Iroquois of North America, a powerful sea turtle, the Great Turtle, once again carries Earth on its back. Because many species of sea turtles have thirteen large scutes on their shells, in these traditions the turtle also represents the lunar calendar with the scutes symbolizing the thirteen full moons in each year. Sea turtles have also found their way into contemporary literature as symbolic beings. In his best-selling novels, the English author Terry Pratchett revisited the image from Hindu mythology of a sea turtle swimming through the universe. His fictional Discworld rests on four elephants (Berilia, Tubul, Great T'Phon, and Jerakeen), which in turn stand on the back of the World Turtle, Great A'Tuin.

On the other hand, despite these forms of worship and even though humans stepped onto life's stage many millions of years later than turtles, people have caused a great deal of harm to sea turtles in a short amount of time. They have pushed sea turtle populations around the world to the brink of extinction. The first real-world interactions between sea turtles and people are well represented in archaeological finds, cultural artifacts, and historical documents. These records prove and tell of a shared history that stretches back to the Stone Age, five thousand years ago. People caught sea turtles as food and crafted tools, religious objects, and jewelry from their shells and bones. Archaeologists have found sea turtle remains in human burial sites around the Mediterranean, the Arabian Peninsula, and North America. In ancient Egypt, the scutes from hawksbill turtle shells—known today as tortoiseshell—were used to fashion rings, bracelets, bowls, knife handles, amulets, and combs.

But it wasn't until colonial times that the sea turtles' story turned into a tragedy, and our food preferences threatened their existence. When seafaring nations conquered tropical and subtropical countries, they also conquered the habitats of most species of sea turtles. Like many other customs around "exotic" foods, the newly acquired knowledge of how to prepare and eat sea turtle meat, especially the meat of green turtles, found its way back to Europe via the colonial overlords. The almost three-hundred-year-long craze for turtle meat that followed led to massive overharvesting and to the almost complete extinction of some species.

Although eating turtle meat in Europe was initially reserved for those in high society, the first Europeans to eat sea turtle meat were anything but upper class. Seafarers took living turtles on board as provisions so they could have fresh meat on long journeys. Their knowledge about the edibility of sea turtles came from the Indigenous inhabitants of Caribbean islands, where large turtles also served as cheap sources of meat for enslaved people in the newly conquered colonies. The West Indian elite enslavers soon acquired a taste for turtle meat themselves and imported turtles to Europe as a luxury dish.

By 1753, there was an established market for sea turtle meat and soup in the British Isles. The *Gentleman's Magazine* reported on large turtles being prepared as menu items in London hotels. Demand for sea turtle meat rose so rapidly over the years that ships plying routes to the West Indies were equipped with wooden pools in which living turtles could be transported. Relatively quickly, turtle soup became a dish for the European elite, who dined on the fruits of colonialism. In the eighteenth and nineteenth centuries, turtle soup was considered a delicacy of the highest order and was a staple on menus in the most exclusive restaurants. For centuries, the soup itself has consisted of a clear broth with meat from green turtles, hence the name that was used for this species in Germany for a long time: *Suppenschildkröte*, "soup turtle."

The soup's later popularity was tied to the increasing importance of the upper middle class toward the end of the nineteenth century. The golden years of the aristocracy were fading, and the newly rich middle class were showing off their power and wealth by buying expensive materials and the finest in food—and suddenly turtle soup was on menus from the Ritz to the *Titanic*. It became so popular in both Europe and North America that it started to be commercially manufactured and sold in cans. It was featured on the shelves of delicatessens as "Clear Green Turtle Soup." From there it made its way into the pantries of the rich, and in later decades it also appeared in regular grocery stores for middle-income households. One of the most successful processors of canned turtle soup was in Key West, Florida. The operation belonged to a Frenchman, Armand Granday. In Germany, a firm founded in 1921 by former ship's cook

Eugen Lacroix, and named after him, became the main supplier of "authentic turtle soup" in a can. After the Second World War, this firm satisfied the country's desire for exotic delicacies and affordable luxuries, profiting from the rapid reconstruction and development of the German economy after the war. In 1959 alone, Lacroix's factory in Frankfurt processed about 250 tonnes (275 tons) of frozen green turtle meat into soup.

Turtle soup and other parts of sea turtles were, however, not only a delicacy for gourmands but were also thought to possess healing properties. Christopher Columbus reported from the Azores in 1498 that lepers bathed in sea turtle blood to alleviate their symptoms. For a long time, seafarers (mistakenly) believed that eating sea turtle meat helped protect them against the common scourge of scurvy, which is caused by a lack of vitamin C. Turtle soup was also the main specialty in the spa-like restaurant and health resort Julien's Restorator, founded in 1793 in Boston as a place where the "infirm in health [and] the convalescent" could find a healthy environment and suitable nourishment.

High demand for sea turtles quickly led to the decimation of populations in some Caribbean colonies as early as the beginning of the eighteenth century, and so efforts were made to limit the catch. This prompted Europeans to widen their search in the nineteenth century over the Western Atlantic to Florida, the Cayman Islands, and other places. Turtlers, small boats that specialized in catching sea turtles, became an important part of the maritime economy of the Western Atlantic. They were a critical part of the development of the trade in sea turtles in which turtles were caught and shipped to North American and European consumers. The massive overharvesting that followed forced the canneries that processed turtle soup to scale back production and finally led to strict regulation of the turtle harvest. With the Endangered Species Act of 1973, it became illegal to kill sea turtles in U.S. waters, and many turtle factories declared bankruptcy.

Since then, turtle soup has practically disappeared from our collective memory as a food item. But just fifty years ago, tourists visiting Florida could order turtle soup, turtle steaks, or even turtle burgers, and until the 1990s, the recipe for turtle soup could be found in one of the most popular cookbooks in the U.S., *The Joy of Cooking*. In Germany, although no laws forbade eating sea turtles, the consumption of turtle soup dropped as environmental awareness increased in the 1970s and people condemning the capture and slaughter of sea turtles became increasingly vocal. Despite these changes, Tengelmann, which was one of the first German supermarkets to remove turtle soup from its shelves, only did so in 1984.

Until fairly recently, then, it was still considered acceptable in Europe and North America to eat sea turtles. And so it's not surprising that sea turtles continue to be eaten in many other parts of the world. A recent study estimates that more that 1.1 million sea turtles have been killed over the past thirty years for human consumption. Knowing what we know today about threats facing sea turtles, it's easy to judge people who eat their meat. But that is a very privileged way of looking at things. Many from the Global North take exception to the cruel and bloody way in which sea turtles destined for people's tables are killed, especially as turtles are threatened with extinction. But for many people indigenous to Africa, Asia, South America, and certain groups of islands, slaughtering sea turtles is no different from slaughtering cows, pigs, and other domestic animals, all of which are killed and eaten in Europe and North America without a second thought.

When I first came to Costa Rica, I was forced to grapple with this issue. In those early years, my world was starkly black and white. In my mind, eating sea turtle meat was evil and that was that. I could not conceive of it simply being part of how people lived their lives. Then I got to know people I liked and respected who went out at night hunting sea turtles and their eggs under the cover of darkness. Over time I had to admit that my opinion was profoundly European and that my perspective was colored by what we in the Western world consider to be right and wrong. Now I realize that when all is said and done, a taboo against eating a specific food reflects a societal norm or a religious edict. It doesn't necessarily say anything about an individual, only about the culture in which they grew up. Even so, my first glimpses into the brutality of killing sea turtles are burned deep into my memory and my soul, and when I think back to them, I relive them as though seeing them for the first time.

I HAVE MY FIRST DIRECT EXPERIENCE with the aftermath of a turtle hunt not long after my arrival in Costa Rica. While out on night patrol, I notice tracks of a green turtle in the sand leading up to the vegetation line. As I see no corresponding track leading back into the water, I happily follow the first one, expecting to catch the female digging her nest or laying her eggs. When I reach the end of the tracks, however, I see no sign of the turtle. I look around using the red light of my headlamp, I stick my head into the dense vegetation, I listen for the sound of crawling or the sound of slapping flippers coming from the forest. I hear nothing but the nightly concert of crickets and frogs. It is as though the female has grown wings and flown away.

I am beyond confused and use my two-way radio to contact a colleague who is patrolling a different section of the beach. In broken Spanglish, he tries to explain to me that there's a rational explanation for the female's disappearance: she has likely been snatched by poachers. I turn on the white light on my headlamp and shine it on my surroundings once again. This time, where the flipper tracks end, I clearly spot human footprints leading off into the jungle. A little farther, at the edge of the forest, there's a fresh trail where something has been dragged. It leads into the trees. The poachers must have first carried the female a short distance to cover her tracks. Then they turned her over onto her back and tied her powerful front flippers to her shell so that she could no longer flap around with them. I follow the trail a few hundred yards into the forest to a dark puddle that has already almost completely soaked into the forest floor. At one spot the ground looks churned up. Footsteps lead farther into the forest, but I stay in the place where someone has clearly been up to no good. Acting on a hunch, I kneel down and begin to dig with my bare hands. It's not long before my fingertips touch something hard and yet flexible that's completely encrusted with sand. I make the hole bigger and pull the object out of the ground with both hands. It is round and slick and keeps slipping from my fingers. Finally, I get a better hold on it. The metallic smell of blood fills my nose and mouth. When I direct the beam from my headlamp onto my hands, I find myself looking straight into the darkly gleaming eyes of a dead green turtle. The poachers have cut off the female's head and left it here. I am overcome with dismay and disbelief, and experience a feeling of horror as I gaze into the empty eyes of this magnificent creature.

A few years later, local rangers ask me to help them identify confiscated sea turtle products. This happens all the time, most often when a tourist or a vigilant local notices sea turtle eggs being sold at one of the weekly markets. We are then brought along to identify the species they belong to. This time, however, we travel in an official car not to a market but to a private house in a poorer part of town. As soon as the car door opens, the humid midday heat of the Caribbean envelops us. A lot of police cars with blue lights are parked in front of a house that looks unusually prosperous for the neighborhood. Officers dressed in black, wearing bulletproof vests and sunglasses and carrying machine guns, are securing the perimeter. Apparently, in the course of a major raid, the police have come across not only a lot of drugs but also sea turtle parts, and they need our expertise so they can catalog everything.

The officer in charge leads us into the garage and warns us that it's not a pretty sight. Right away we are hit with the smell of spilled blood. The garage is unlit and my eyes take a moment to adjust to the low light. I notice some white bags along the wall; they are filled to the top with sea turtle eggs. A side door opens and a little more daylight spills into the garage. Then a picture of hell emerges in front of me. A huge stinking pool of blood almost completely covers the garage floor. Hundreds of flies, startled by our presence, fly up and buzz around in the air. In the middle of all this lies the mutilated body of a green turtle. Her four flippers have been hacked off and her plastron has been sliced open along the outer edge. The butchers must have been disturbed. The enormous pectoral muscles have already been exposed. On a long table, already cut into portions, lies the flesh of at least four more green turtles.

As the scene gradually registers, I suddenly taste salt. Tears are streaming uncontrollably down my face. Tears of grief for these wonderful animals, which have made it to adulthood despite all the dangers they have faced, only to end up like this. But also tears of rage over the agonizing way in which these turtles had to die.

Suddenly we hear a raspy breath. One of the two rangers turns on his flashlight and we look around. Right in the back of the garage we find five more motionless green turtles lying on their backs, their flippers tightly bound. Judging by the breath we heard, at least one of them is still alive. We rush over and examine each turtle for signs of life. Sea turtles can hold their breath for a long time, which sometimes makes it difficult to tell if they are dead or alive. I carefully touch the turtles' eyes and eyelids. Even the weakest turtle always reacts with a slight twitch. And indeed: all five turtles are still living. Unfortunately, they have holes in their shells where arrows from harpoons have punctured them. Much like a broken arm bone, a turtle's shell can heal. However, it needs to be treated and monitored by a specialized vet in a sufficiently large and suitably equipped facility where the turtle can be looked after as it heals. The biggest danger when there are holes in the shell is that bacteria will get into the body cavity and lead to infections.

My two companions take turns feverishly phoning their superiors and different wildlife rehabilitation centers and veterinarians. Unfortunately, at this time in Costa Rica, there are no rehabilitation centers capable of taking in five adult turtles at once, nor are there any veterinarians that specialize in sea turtles. Our options, therefore, are limited. We could try to end the turtles' suffering by killing them. However, that's not as easy as it sounds because the brains of reptiles can function for a long time without oxygen. A few years earlier we had taken a turtle that had been attacked by dogs to a regular dog and cat veterinarian. He had no idea how to euthanize a sea turtle. We had to look up the answer on the internet, but the doctor did not have a sufficient supply of the anesthetic on hand for the suggested method, which was to give the turtle an overdose. In the end, he took a drill and perforated the skull to puncture the brain.

This drastic method demonstrates just how incredibly tough sea turtles are; they are almost indestructible. Over the years I have seen females with injuries that looked as though they should have been fatal, but that didn't stop them from coming to the beach to lay their eggs. I have seen flippers half amputated by fishing line rotting on their bodies. I have seen turtles flayed by boat propellers and the damage done by enormous shark bites. Sometimes we witnessed unbelievable recoveries over the course of just a few weeks until ugly scars and missing flippers were the only signs of what had happened. We absolutely want to give these five turtles a second chance, and so the rangers go out and buy synthetic resin and strips of fiberglass at the nearest hardware store and iodine at the drug store. We do what we can to disinfect the wounds with the iodine and create a makeshift patch over the holes with the resin and fiberglass. Then we set the turtles free. We don't have high hopes that they will survive, but at least this way they stand a chance.

These experiences have never left me. They and many others are fixtures in my recurrent nightmares and are the reason I am generally not a fan of eating meat. But sea turtles are killed not only for their meat. In fact, in some species, according to people who know about these things, the taste leaves something to be desired. In the case of hawksbills, a meal of sea turtle meat can even be life threatening. In many regions of the world, hawksbills live mainly off sponges, one of the oldest multicellular forms of life. Sponges have no organs, no muscles, no neurons, and no brain. They are usually firmly attached to the substrate on which they grow and feed themselves by filtering particles from the water. What is most fascinating about them is their defense mechanism. As they cannot flee from predators, they use toxins. Some of the strongest natural toxins have been found in sponges. However, these poisons appear to be useless against hawksbills, which eat the sponges and accumulate the toxins in their own flesh. If a person were then to eat the turtle, they could die of what is known as sea turtle meat poisoning or chelonitoxism, especially if the meat is not cooked long enough. The meat of a green turtle can also poison people and other animals if the turtle has eaten a large amount of the toxic algae that form a red tide. These single-celled organisms produce a potent nerve toxin, brevetoxin, that accumulates in muscle meat, and if enough accumulates it can also lead to the death of the turtle itself.

Another product from sea turtles is oil, which can be rendered from the thick fat layer of leatherback turtles in particular. Sea turtle oil is still used as fuel for oil lamps, as a waterproofing material for boats, and as a remedy for respiratory illnesses. And enormous slaughterhouses in Mexico that were in operation up until the 1990s specialized in producing boot leather from olive ridley turtles; they were almost single-handedly responsible for wiping out local populations.

Sea turtles are not aggressive and have no way of protecting themselves against people. And that is why they need places where they can carry out the most important activities for their survival undisturbed: eating, reproducing, and resting. In scientific terms, these places are called critical habitats, and you can imagine them as the equivalent of a house. The turtles' feeding grounds in this case would be like our kitchen, breeding grounds would be like our bedroom, and places to rest would be like a hammock in the backyard.

Disturbances in these habitats have grave consequences not only for the ability of individual sea turtles to survive but for the whole species. It would be like someone throwing the bed out of the bedroom or removing the fridge from the kitchen, jumping up and down on the mattress while we were trying to have a few romantic hours with our partner, or dancing on our kitchen table when we wanted to eat. It would be even worse not to be safe in our own home because someone is perhaps lurking in a dark corner waiting to attack and kill us while we go to the fridge unaware of the danger. And yet many sea turtles fall victim to people right in their breeding and feeding grounds, and unfortunately hunting is not the only threat they face. Our lifestyle and the way we grow food on land are huge threats to ocean life and therefore also to sea turtles. Fertilizers and pesticides from agriculture find their way into our oceans and remain suspended in the water, invisible to the naked eye. It doesn't matter if we live by the ocean or not, because all rivers eventually lead to the sea. These chemicals poison the drinking and bathing water in the turtles' homes—with dramatic consequences.

Scientists have been puzzling over a mysterious sea turtle disease for decades. Since the late nineteenth century, there have been more and more reports of a kind of sea turtle cancer that causes cauliflower-like tumors on their skin and also in their eyes, mouths, and internal organs. Some sea turtles exhibit only a mild form of this disease, whereas others develop numerous enormous tumors that can lead to extreme weakness and eventually death. This disease, first reported in the scientific literature in 1938 and dubbed turtle fibropapillomatosis, is caused by a herpes virus known as chelonid alphaherpesvirus-5 (ChHV5). It gained international attention in the 1980s when it reached epidemic proportions in regions such as Florida and Hawaii. A striking feature of the disease is that it mostly attacks populations living in waters heavily polluted with fertilizers, such as the coastal waters of Hawaii, where the land has been scarred by pineapple plantations.

Fibropapillomatosis has now been found in all seven species of sea turtles, but green turtles are most often and most severely affected. Tumors appear less often and in milder forms in the other species. In places where the disease is relatively common, more than half the green turtles examined developed the disease within a year of arriving in their feeding grounds.

We know that fibropapillomatosis can be transmitted from one turtle to another, but we don't yet understand how exactly they get infected. One hypothesis is that other marine animals play a role. The herpes virus that causes the disease has been found in parasitic leeches, which attach to the skin of sea turtles and suck their blood; it has also been found in the mouths of cleaner wrasses. The virus survives in sea water and so could travel through the water from sea turtle to sea turtle. Epidemiological studies indicate that the virus is acquired when young sea turtles arrive in coastal feeding grounds after spending their first years out in the open ocean.

In 2010 researchers discovered that turtles that live and feed in parts of the ocean with higher nitrogen concentrations—for example, near river estuaries, which carry lots of fertilizers into the sea—are more likely to suffer from the disease. To further investigate this link, they researched nitrogen accumulation in the algae the turtles feed on. These algae convert excess nitrogen into an amino acid called arginine. Turtles that suffer from fibropapillomatosis have higher levels of arginine in their systems than healthy turtles, and the algae the diseased turtles are eating also contain high concentrations of this amino acid. We also know that arginine promotes the growth of herpes viruses, including the human herpes simplex virus. People like me who occasionally suffer from cold sores likely know that eating foods with a high arginine content, such as walnuts, makes an outbreak more likely. Interestingly, more than two-thirds of the human population carries the virus but not all will develop the typical symptoms of this disease. Mostly it affects those with a weakened immune system—for example, people who suffer from chronic stress. Fibropapillomatosis in turtles seems to show a similar pattern. It could be, therefore, that most turtles carry the virus, but the disease only manifests itself in individuals that find themselves in waters containing a lot of fertilizers and whose populations are under intense environmental pressures.

Feeding grounds are retreats for sea turtles—the places where they grow and recharge their supplies of energy. They spend most of their lives feeding, because this fuels the activities that sustain their populations: migrating long distances to breeding and nesting grounds, producing eggs and sperm, mating, and laying eggs. To be sufficiently prepared, males as well as females eat as much as they can in their feeding grounds before they set off on their long journeys to the places where they will mate. This is precisely why it is so important that we offer them more protection there.

LOGGERHEAD TURTLE

Caretta caretta

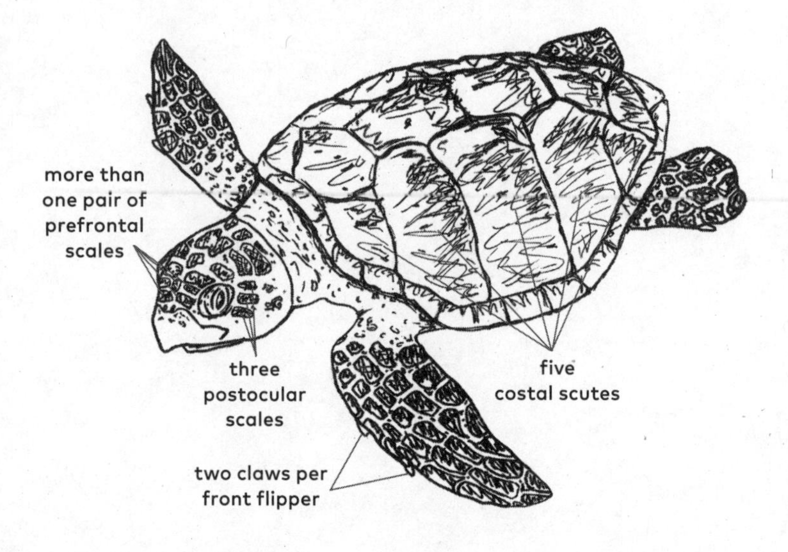

IUCN STATUS: VU—Vulnerable

CARAPACE LENGTH: 80–110 cm (31–43 in.)

WEIGHT: 70–170 kg (154–375 lb.)

MAIN DIET: Hard-shelled invertebrates

AGE AT MATURITY: 28–33 years

REMIGRATION INTERVAL: 2–4 years

INTER-NESTING INTERVAL: 15 days

NUMBER OF CLUTCHES PER NESTING SEASON: 4–7

NUMBER OF EGGS PER CLUTCH: 115

INCUBATION TIME: 50–60 days

GOOD TO KNOW: Huge head with extra-strong jaws (bite force 2,100 newtons/472 pounds-force); only species to nest in temperate climate zones

– 5 –

THE VOYAGES OF SEA TURTLES

MY ASSISTANT MACDONAL and I are floating on the surface poised for what will happen next. It's a sunny day, the water is deep blue, and visibility is good. Even so, we cannot see the bottom below us. We have just spotted a pair of olive ridley turtles in the act of lovemaking and are counting the seconds until the male lets go of the female. When he finally slides off her back, I dive, closely followed by Macdo. I take a couple of powerful strokes and before the female realizes what is happening, I am under and behind her and grabbing her shell as she prepares to dive. She uses all her strength to try to swim away and drags me a short distance through the water. I manage to direct her toward the surface to get her chest and front flippers out of the water so she can no longer propel herself away from me.

After my exertion of swimming fast and wrestling with the turtle underwater, I need to take a couple of quick breaths. Carefully, I adjust my grip on the female's shell, holding on more tightly so she can't escape, and look around for our boat. It's bobbing up and down on the waves about 50 meters (55 yards) away. I can make out the figures of our captain, Santiago, and my colleagues and assistants Brie, MJ, and Marcus. I notice that Macdo has already swum off in their direction now he's seen that my hunt has met with success. With the female firmly in my grip, her flippers flailing wildly and narrowly missing giving me one slap in the face after another, I, too, swim back to the boat, using only my flippers.

Marcus and MJ are already waiting for the turtle. With what strength I have left in my legs, I flutter-kick and lift her a few more inches out of the water, and the two of them take her from me and pull her into the boat. To keep the turtle's stress levels to an absolute minimum, they cover her eyes with a dark, damp cloth. Once she is safely on board, Macdo and I heave ourselves up over the railing on the opposite side and into the boat. While I attempt to catch my breath and dry off a bit before I drip all over our equipment, Marcus holds the turtle securely by the shell on one of the boat's bench seats and MJ begins a preliminary inspection.

I glance down at my thigh and see bloody scratches from the barnacles and other crustaceans living on the boat's hull. In the heat of the moment, I hadn't noticed that I had scratched myself to the point of drawing blood. Catching olive ridley turtles in Costa Rica is part of my PhD research into their lives out in the ocean. I am particularly interested in where they hang out during nesting season between laying clutches of eggs and where they travel after nesting is over. To help me with this, I am attaching transmitters to a small number of females so we can track them during the nesting season and follow them afterward.

It's not easy to attach a transmitter to a turtle because the attachment must withstand wet conditions and the immense drag in the water—preferably for months. Several methods have been tried in the past, ranging from a backpack-like harness held in place by a metal ring that slowly rusts through, to a floating transmitter that the turtle would drag behind itself, secured with a length of fishing line looped through a hole in the rim of the carapace. The first method, however, led to the premature loss of the transmitter if the turtle lost a lot of weight, or injuries in the soft flesh above the front flippers if the turtle put on a lot of weight. The second method reduced drag but the transmitter often tore off when the turtle swam through rocky areas. Over the years, the best method—for adult sea turtles, at least—proved to be gluing the transmitter to the carapace using cold-setting epoxy resin designed to cure in humid environments.

After MJ finishes gathering the initial data and declares that the female is an excellent candidate for our study, we motor back toward the harbor so we can attach the transmitter on land in the shade. When we arrive, MJ and Marcus carry the turtle to the beach and lay her on a mat under a large, spreading mango tree. Marcus positions himself behind the turtle and holds on tightly to her front flippers, while MJ helps restrain the turtle from the front. We are about to begin the tricky part of the operation, and we want the turtle to move as little as possible. We restrain the turtle manually. Other projects use crates to immobilize turtles, especially the larger species, to make sure they don't injure either themselves or others. If we are attaching a transmitter during the day, we do everything in the shade—in our case under a tree—so the turtle doesn't overheat.

Working with epoxy resin is an extremely sticky—and tricky!—business. Depending on the resin, the consistency ranges from liquid to modeling clay. No matter what form it takes, you have only a few minutes to mold it because the moment it dries it's almost impossible to remove. I must, therefore, work carefully and meticulously but also quickly to ensure that the resin lands in precisely the right spot before it hardens. Unfortunately, when you're working with a struggling turtle, that doesn't always happen. Over the years, I have accessorized quite a few pieces of clothing and research equipment with resin and have walked around for weeks with a mixture of sand and resin in my hair, on my face, and on my hands until my resin-decorated skin flaked off and my nails grew out. A few colleagues have even managed to stick themselves to their research subjects. Not a particularly attractive thought if you happen to be researching crocodiles rather than sea turtles.

Before I attach the transmitter to the shell, I must first remove any slippery microalgae or other organisms adhering to it. To do this, I use a spatula, a stiff brush, a pot scrubber, and fresh water, all of which we have brought along with us for just this purpose. After multiple rinses, the last traces of microalgae are washed away, and I finish cleaning using a rag dipped in acetone. Next, I roughen the shell with sandpaper to give the epoxy resin more surface area to adhere to. Finally, the preparations are complete and it's time for the transmitter. I spread sticky epoxy resin on the prepared section of shell and firmly attach the transmitter. I then apply additional layers of resin around and over the transmitter to create as tight a seal as I can to protect it from saltwater damage.

During my first season working with transmitters, I still used the somewhat outdated technique of attaching them using fiberglass strips and synthetic resin that cures when you mix it with a hardening agent. Unfortunately, this method doesn't work particularly well in the tropics. High humidity and warm temperatures slow the curing process. Theoretically, you can shorten the time by increasing the amount of hardener in the mixture, but that would in turn increase the temperature of the curing resin so much that the turtle might sustain lasting damage to its lungs, which are located right underneath the part of the carapace we attach the transmitters to. Therefore, it's always better to just wait patiently until the resin has cured and the turtle can be released back into the wild. Using newer epoxies, that usually takes no longer than one hour.

But sometimes it feels as though the curing process drags on for an eternity. I remember seemingly endless nights on the beach, after weeks without enough sleep, with one of my assistants restraining the turtle all night long. For the assistants, these nights meant spending hours in an uncomfortable position, often while being besieged by thousands of tiny biting sandflies that attacked every inch of exposed skin while their victim had no free hand to fight them off. Meanwhile, the rest of us would sit around trying desperately not to fall asleep as I tested the consistency of the resin every five minutes. Many of these nights ended with the horizon gradually becoming lighter as it picked up the first rays of sun reflecting off the mountains, bathing those of us on the beach in an otherworldly glow. It looked as though a unicorn had galloped across the sky, leaving behind an aerial trail in the most dazzling shades of red, orange, purple, and green—perhaps to help us more quickly forget the ordeals of the previous nights and weeks. When it was finally time to remove the black cloth from the turtle's eyes and send it on its way, we were intoxicated by lack of sleep and celebrated by yelling rather than singing along to the songs on our season's playlist as we drove back to camp.

It's not necessarily faster or easier to attach transmitters during the day. One day, when we left for the ocean at dawn, the sky was still clear, but by the time we were back on land with a turtle, thick clouds were gathering in the distance threatening rain and stormy weather. The epoxy resin refused to harden in the humidity, and the black clouds above us became increasingly threatening until suddenly, even though it was still early afternoon, it felt as though evening was pressing down upon us.

By the time the resin finally dried, we could hear the rumble of thunder in the distance. Now, however, we still had to return the turtle to the place where we caught her. Under normal circumstances, that would be a twenty-minute boat trip there and back, but the approaching storm was now whipping the previously gentle waves into a lashing spray. We were at the edge of a harbor protected by cliffs, but I didn't want to release the turtle here because I was worried that she might not find her way back out into the open ocean or that she might get injured by the propeller of a fishing boat. I quickly put my students into the shelter of my car and carried the sea turtle with the help of my captain, Santiago, to his small fishing boat. He had dragged the boat out of the water and up the sand, but the waves were now pulling and tugging at it as they tried to sweep it off the beach—and now we wanted this little walnut shell of a boat to carry us into the foaming sea? The wind was whistling past our ears and we had to shout at each other to make ourselves heard. I felt anything but relaxed as I jumped in the boat to join the turtle and held her firmly with both hands. As Santiago was the picture of calm, I silently gave myself a talking-to to buck myself up and hoped that my captain knew what he was doing.

With a single shove, Santiago pushed our boat out into the water and then jumped in. With lightning speed, he started the motor and gunned it out into the open ocean far from the beach and the rocky cliffs. He took the increasingly tempestuous waves head on so we didn't capsize. At times we were flying above rather than gliding through the waves and we crashed down hard on their back side. Then the floodgates opened and we were soaked by torrential tropical rain. Lightning flashed directly above our heads, followed seconds later by deafening thunderclaps. At the bottom of the wildly tossing boat, I cowered over the turtle with my eyes shut and prayed.

Once we were out of the worst of the waves, I dared to open my eyes. My gaze fell on Santiago, lit up by lightning. He had a wide grin on his face.

"Now that's what I call a storm," he shouted.

"Aren't you scared?" I asked.

"The Lord giveth and the Lord taketh away. If today is the day my time is up, then so be it."

At that moment, I wished I could share his serene acceptance of what the future might bring, as truth be told my heart had jumped into my throat and I was close to panic.

As soon as we were a couple of hundred yards out into the open ocean, we lifted the turtle to the railing and carefully let her slip into the water. Travel safe, I told her in my thoughts. I watched as first she and then the antenna on her transmitter disappeared into the roiling gray-green waves. Then we beat it back to the harbor.

SEA TURTLES SPEND most of their lives at sea. After they have swum out from the beach as babies into the open currents, as a general rule the only ones to leave the ocean after that are females, who return to the beaches of their birth to lay eggs. As they mature, young sea turtles travel from one feeding ground to the next until they finally decide on an area where they will remain as adults. Olive ridleys are the exception and are known for their continued nomadic wanderings through the oceans in search of food.

Both males and females of all species return to the area where they hatched to mate—and for the females to lay their eggs. As a result, adults undertake long migrations back and forth between their feeding grounds and nesting beaches, for years. Hard-shelled species cover an average of 1,000 to 3,000 kilometers (620–1,860 miles) on these trips; leatherbacks commonly swim 12,000 kilometers (7,450 miles) or more from their temperate feeding grounds to their tropical nesting beaches. The longest recorded travels of individual sea turtles are even more impressive. A female green turtle completed a roughly 4,000-kilometer (2,500-mile) journey from the Chagos Islands in the Indian Ocean to the coast of Somalia in East Africa. The longest documented journey by a leatherback turtle, which swam from the coastal waters of Indonesia to the Oregon coast in the United States, was 20,558 kilometers (12,774 miles). The record holder, however, is Yoshi, a loggerhead turtle. She was rescued as a youngster from a Japanese fishing boat and raised in the Two Oceans Aquarium in Cape Town,

South Africa. When she was released eighteen years later, it took her close to three years to swim an astounding 40,011 kilometers (24,862 miles), crossing an entire ocean to reach what were presumably her home waters in Western Australia.

Sea turtles are perfectly equipped for long-distance swimming. Over the course of evolution, their shells and flippers have been fine-tuned for marathon swims. They lost the ability other turtles have to withdraw their limbs and head into their shells. Their shells became more streamlined and the massive pectoral muscles that power their front flippers took up the last remaining space inside. Sea turtles can also store enormous reserves of fat, which provide the energy they need to complete their odysseys.

Sea turtles are not only skilled swimmers but also expert navigators. One of their most remarkable—and most mysterious—talents is their ability to return to the areas they originate from. We don't yet have a detailed picture of how they find their way back to the beach where they took their first breath after being away for years, even when they, like Yoshi, have been completely diverted from their usual travel routes. One thing we do know—and have known for some time—is that, like migratory birds, they can sense Earth's magnetic field. They can orient themselves by the intensity of this geomagnetic field, as well as by the angle of inclination of the magnetic field lines in relation to Earth's surface. Both change in a predictable pattern around the globe so that, generally speaking, every region of the ocean has a unique magnetic signature. Presumably sea turtles can use these signatures as a kind of compass or GPS, allowing them to use both latitude and longitude to swim in the right direction. Recent studies suggest that turtles learn the characteristic magnetic signature of their home beach through what is known as geomagnetic imprinting. Preliminary data are giving us the first evidence that even sea turtle embryos do not move randomly in their eggs but apparently use Earth's magnetic field to orient themselves.

Earth's magnetic field, however, cannot be the only directional guide sea turtles use. In experiments where their geomagnetic sense was impaired, test turtles were still able to navigate and eventually arrive at their destination. Sea turtles must, therefore, also use other sources of information, like the position of the sun and other stars, the direction of the waves, or olfactory indicators in the water, in much the same way that blind people are often able to orient themselves using nonvisual clues. We don't yet know how exactly that works, and there is a pressing need for more research in this area.

It's extremely difficult to follow sea turtles across oceans and fathom the details of their aquatic lives. The first studies relied on simple mark-recapture techniques. Turtles were outfitted with numbered metal tags on their flippers and a finder's fee was offered for every tag found. Mostly they were found by fishers, who caught the tagged turtles and returned the metal tags to the relevant university or research station. Using this method, even back in the 1950s and 1960s scientists knew, for example, that nesting females did not feed near their nesting beaches between nesting seasons. Even so, they had no idea just how far sea turtles actually traveled.

Dr. Archie Carr, the great-grandfather of all sea turtle biologists, was also interested in sea turtle migration and used what seems to us today a somewhat barbaric method of tracking nesting females in the ocean after they had laid their eggs. He tied them to helium-filled weather balloons that they then dragged through the water. The balloons were clearly visible at the surface, and Dr. Carr could follow them by boat and gather data about the routes the females were taking. This crude technique didn't stand the test of time because the balloons severely restricted the females' ability to swim and make deep dives. In addition, scientists could only follow turtles for relatively short distances before losing sight of the balloons. I would also bet that these turtles did not behave as they would have had they not been tethered to balloons. But these early studies clearly demonstrate that even back then there was great interest in turtles' lives in the oceans and in the journeys they undertake.

The arrival of VHF radio transmitters increased the radius of possible research to about 15 kilometers (9 miles). I still sometimes use these transmitters in my studies. They allow researchers to track animals from a greater distance and still locate them again and again. These devices send a signal at very high frequencies in the 137 to 225 hertz range. The signals travel to a directional antenna, and a receiver converts them into sound. Depending on the brand, the sound is a knock or a pulse that gets louder the closer the receiving antenna is to the transmitter. This allows you to home in on an animal's position by adjusting the position of the antenna.

The transmitter itself is a small box with an antenna and weighs less than a package of gummy bears. The rule of thumb is that the transmitter should always be lighter than

5 percent of the tracked animal's body weight. A transmitter is usually attached at the highest point of a turtle's carapace, which is above the water when the turtle surfaces for extended periods. Of course, a transmitter affects the turtle's hydrodynamics the way any object attached to its shell increases water resistance, including gooseneck and other barnacles, which love to grow there naturally. More resistance means the turtle uses more energy when it swims, but that doesn't make much difference to a healthy animal. Therefore, for our studies we choose only animals in excellent condition; they can easily adapt to a bit more weight.

The VHF radio system, however, has its limitations when used with animals that live in a marine environment. On land, as long as the animal is within receiving range, the sound is always audible and all the researcher has to do is walk in the direction of the transmitter. With a sea turtle in the water, however, you have to wait until the turtle surfaces, and then you have to try to drive the boat in the right direction. I have to say that we don't often succeed in finding and catching a specific sea turtle when it's in the water. However, during some of the legendary mass nestings of olive ridley turtles—the arribadas—transmitters allowed us to find the proverbial needle in a haystack. Among thousands of nesting females on a miles-long beach, we were successful in locating our tagged turtles, the ones we wanted to take blood samples from multiple times for our study.

With advances in technology and the advent of global systems for navigation and data transmission via satellites, there has been a corresponding increase in ways to research animals that travel long distances. Satellite navigation systems generally use the Doppler method to determine position. Satellites receive the signal from the transmitter and then relay it to a station on the ground. Receiving signals via a series of satellites (or multiple time-delayed signals received by the same satellite), and using an algorithm that takes into account the curvature of Earth, pinpoints the location of an animal. The precision of this method depends on the number of satellites used and the duration of the signal. The latter is an especially critical factor when working with marine animals. They usually come to the surface only briefly, which means that many of the pinpointed locations end up being off by many hundreds of yards or more.

The satellite transmitters we use for sea turtles look very similar to VHF radio transmitters except they have what is known as a saltwater switch. This switch recognizes whether the transmitter is above or below water and controls a circuit that initiates or interrupts satellite transmission. If a transmitter tried to establish contact while underwater, it would waste energy and shorten the life of its battery. Generally, batteries last several years, and if a transmitter stops working it's usually for some other reason. The antenna might snap off, the saltwater switch might become overgrown by barnacles and cease to function, or the transmitter might fall off the carapace because the attachment fails.

ONE NIGHT I'M SEARCHING for the perfect candidate for one of my satellite transmitters, when I see a female olive ridley struggling up the beach laboriously dragging a tangled mass of fishing net. A strand of net has wrapped itself around her right hind flipper, cutting through flesh and bone and almost completely amputating the flipper, which is now hanging on by a single ligament and is being dragged up the beach along with the ball of net. The flipper is bloated and grotesquely distorted and giving off an awful stench. The poor female must have been in agony for weeks. I fish my Swiss Army knife out of my fanny pack and start my rescue mission by cutting the main strand of net attaching the tangled mess to the turtle. Then I inspect the flipper more closely. Even though the limb looks dead, the turtle flinches when I touch it gently. The flipper still wrapped in the rest of the line is definitely beyond saving, but the exposed surface of the stump it has left behind looks surprisingly good under the circumstances. No swelling or pus, and there are signs that the injury is beginning to heal.

What should I do? Ideally in these circumstances, you would inform the authorities and hope there's a good rehabilitation center around where the turtle can be treated with antibiotics and, if necessary, an amputation. However, I am in the middle of nowhere, without cell phone reception and several hours from the nearest veterinarian who might have even the slightest idea of what to do with a sea turtle. I don't have a way to get the turtle off the beach anyway, let alone transport it to another part of the country. And at this time of night I can't reach any of the local government officials responsible for sea turtles because they are all sleeping soundly in their beds. Quite apart from that, in the past I have often been told, quite curtly, that it's best to let nature take its course...

And so it's up to me to find another way to save this turtle's life, or at least do what I can to relieve its suffering.

When you find a sea turtle entangled in fishing line or in a net, the last thing you want to do is to cut the line or net wrapped directly around a limb. There's often a buildup of bacteria in the limb because circulation has been cut off and the tissue is slowly beginning to rot. If you suddenly cut the strand of netting and restore circulation, bacteria could be flushed into the rest of the turtle's body. I must now decide whether to be satisfied with cutting away the heavy tangle of net or whether I can do more to help the female. A careful tug on the ligament shows me the female still has some sensation there. Cutting through it with my pathetic little pocketknife is out of the question as it would only subject her to more needless pain. I quickly decide to cut a strand from the tangled mass of netting and bind it tightly around the ligament close to the stump. I am hoping that will completely sever the remaining nerves so that the flipper will soon get torn off or fall off on its own.

Despite the dreadful state of her flipper, the female continues to crawl up the beach looking for a place to dig her nest. The indestructability of sea turtles never ceases to amaze me. I look down to the bulk of fishing net in my hand. Judging from its condition it must have been drifting in the ocean for quite a while, with or without the female entangled in it. I am constantly horrified by the dangers sea turtles face on their long journeys.

As this story illustrates, apart from the natural enemies that lie in wait for small turtles behind every piece of driftwood, there are numerous human-made threats that can be fatal for adults as well. Foremost among these is industrial fishing. Enormous commercial fleets can empty whole sections of the ocean of fish within a few weeks. Their nets and long lines stretch for miles and snare nontarget species as bycatch, including whales, dolphins, and also sea turtles, all of which have to come to the surface to breathe. If they are lucky and the nets and lines are checked often enough, the unwanted bycatch can be freed—often roughly and with a few injuries but otherwise alive and well. Unfortunately, that is the exception rather than the rule, and often getting caught in a net is fatal. It's an unbelievably stressful situation for the animals. As they frantically swim around and try—and fail—to free themselves, they use more oxygen than usual, sustain grave injuries, and eventually die an agonizing death by drowning that takes hours. An estimated 500,000 sea turtles die each year as bycatch. When fishing nets get damaged, are no longer needed, or break loose, they are simply left in the ocean, where they drift as ghost nets, catching everything that gets too close to them. We don't know how many more sea turtles get tangled up and die in these discarded or lost nets, but the number is likely just as high.

And that leads us directly to pollution of the oceans, another danger that claims the lives of thousands of turtles every year. Apart from the invisible pollution caused by fertilizers and pesticides that I've already mentioned, petroleum products and plastics are two other main threats. Somewhat surprisingly, most of the oil pollution stems from oil and petroleum derivatives carried into our oceans by rivers. Regular shipping traffic is another major source as vessels discharge oil both legally and illegally. Oil slicks, concentrated pollution that occurs after an accident at sea, happen more often than you would think. From 2011 to 2021, there were a total of seventy-two accidents that resulted in oil spills. Many of these catastrophes are not widely reported in the media, especially when they happen in economically underdeveloped countries. In May 2021, for example, a container ship caught fire off the coast of Sri Lanka and sank, carrying 25 tonnes (28 tons) of highly toxic nitric acid, 350 tonnes (385 tons) of heating oil, and 1,680 tonnes (1,850 tons) of plastic pellets, as well as other chemicals and cosmetic products. A few weeks later, hundreds of sea turtles and other marine animals washed up dead on land. And yet to this day many people have not even heard of this catastrophe.

Sea turtles can ingest oil particles, or they might feed on prey animals and plants that have been exposed to oil. Ingestion can cause gastrointestinal irritation, ulcers, internal bleeding, diarrhea, and digestive problems, all of which can affect the turtles' ability to eat and process food, leading to decreased health, reproductive success, and chances of survival. Oil and the chemicals used to disperse spills can irritate or burn sea turtles' skin on contact. Damaged skin can lead to severe infections and the chemicals can be absorbed through the skin, leading to liver and kidney damage, anemia, a weakened immune system, and reproductive disorders. In extreme cases, turtles die. Embryos in sea turtle eggs on beaches affected by oil slicks grow more slowly, which results in lower hatching rates and developmental problems. Turtles might also inhale the volatile organic substances that evaporate from oil floating on the surface, which can lead to irritation and inflammation of the airways, emphysema, and pneumonia.

Oil pollution can also have indirect effects by altering the ecology and behavior of sea turtles—for example, when turtles have to spend more time finding food and thus waste energy, or when they have to endure other disruptions in their natural life cycles. Although research into oil catastrophes has shed more light on the probable consequences of oil pollution, we still don't know the full implications, especially when it comes to long-term damage to affected ecosystems or the physical effects on animals exposed to toxic substances for years.

Another bane of our existence is that miracle of modern engineering: plastic. The term covers a wide range of synthetic or semisynthetic materials. Most synthetic plastics are made from petroleum and therefore contribute to the climate crisis. The success of these products lies in their malleability and versatility. Ironically, early plastics were used to replace tortoiseshell, the similarly malleable and versatile shell of the hawksbill turtle.

Just six types of plastic account for about 90 percent of global plastic production (about 350 million tonnes or 386 million tons per year). We see them regularly in many forms in our day-to-day lives, from sandwich bags to toothbrushes. And we are using more every year. In 2020 alone, we produced 400 million tonnes (440 million tons) of plastic, and global production is expected to rise annually. Estimates suggest that by 2050 global production will exceed 1,100 million tonnes (1,210 tons) a year. The dilemma is that plastic has allowed us to make huge advances in modern technology, research, and medicine, and plastic products probably prevented the extinction of the hawksbill turtle. At the same time, plastic has a more than problematic ecological footprint and poses health risks for both humans and animals. Proper disposal and management is an enormous challenge because plastic products last for centuries in landfills and in the environment.

We cannot recycle ourselves out of the plastic crisis either, as only three of the most-used plastics can be recycled. The message from plastic producers that better recycling will solve the issue of plastic waste pushes responsibility onto consumers when it should really lie with the manufacturers. Moreover, many countries in the Global North send their dirty unsorted plastic to Africa and southeastern Asia, where people must slog through mountains of garbage to find anything that is actually recyclable. The rest of the garbage, of course, remains in these countries, where landfills are far less protected from rain, floods, and storms. And this is why 80 percent of the plastic in the oceans has its origins on land. Every day, rivers carry enormous quantities of garbage toward the ocean—including garbage we have disposed of responsibly. Recent calculations suggest that every year 11 million tonnes (12 million tons) of plastic enter the ocean, adding to the estimated 200 million tonnes (220 million tons) already circulating there.

Entanglements in plastic are not the only way this pollutant can cause a sea turtle's demise. To a turtle, a plastic bag looks like a jellyfish and it thinks it's food, and new studies show that plastic even smells like food. Plastic has been found in the stomachs of all seven species. It can harm or completely block a turtle's intestinal tract. In the latter case, the turtle often dies a slow, agonizing death from starvation because it can no longer ingest food or because food it does ingest cannot be absorbed properly. This is something I encounter often.

One time when I was working with a sea turtle that came onto the beach to nest, I noticed that something wasn't quite right. After the turtle had dug out her egg chamber, she remained in the same position for half an hour but nothing happened. The female appeared to be pressing down, but no eggs appeared. As I was lying on my stomach behind her, holding the bag for the eggs under her tail, I reached out and gently felt for her cloacal opening. Normally turtles really don't like it when you touch them there, and it's not something I ever do under normal circumstances. What I felt, however, was not the regular opening; there was something soft hanging out of the cloaca. I tugged at it carefully and it gave way. Little by little, I pulled it out of the cloaca. When I was done, what I had in my hand was a black garbage bag that had seen better days. Once it was out of the way, the eggs finally started to drop. After the female ate the bag, it must have passed through her entire intestinal tract and gotten stuck inside the cloaca. This turtle was lucky. Many others suffer a much worse fate.

Another threat, one that has not received nearly enough scientific attention, is the effect of microplastic. The problem with plastic is that it isn't biodegradable. Instead of decomposing into harmless components, it degrades, which means that, as it ages, ultraviolet rays and wave action break it down into smaller and smaller pieces. When the particles are less than 5 millimeters (0.2 inches) in diameter, we call it microplastic. We are already finding microplastic everywhere in the environment: on the highest mountaintops and in the deepest ocean trenches; in our drinking water, our food, and our bodies; in our lungs, in our feces, and even in the placentas of pregnant women. In fact, one study claims that we ingest about a credit card's worth of microplastic a week. The main sources are clothes made from synthetic fabrics, which lose plastic fibers with every wash. Our water-treatment systems are not capable of filtering this microplastic out to prevent it from entering our waterways. Our food packaging also sheds plastic. Even these small plastic particles, which usually don't cause any visible physical damage, can lead to reduced biological fitness in sea turtles. Many of the additives used in manufacturing plastic and many of the contaminants that accumulate on the surface of plastic objects over time are highly toxic and interfere with hormones, acting as endocrine disruptors. A recent study has shown that just fourteen ingested plastic particles are enough to raise the likelihood that a turtle will die by 50 percent.

The general threat plastic poses to marine animals has been known for a long time. The first report was published in 1981. It documents the presence of plastic in the stomachs of seven leatherback turtles that were examined after they died. The necropsies it reports on date as far back as 1970. Since then, scores of scientific papers have documented exactly how and to what extent plastic harms marine life. The scientific statistics are frightening but also somehow abstract. It is not numbers that push us to think about the problem and make changes, but stories about the fate of individuals.

THE MOST INCREDIBLE STORY that could probably be told here is that of the sea turtle with a plastic straw up its nose. This event, which I filmed in August 2015—or rather the visceral public reaction to this event—is the main reason many people know who I am.

On that fateful day I was traveling by boat several miles off the Pacific coast of Costa Rica gathering data for my

PhD dissertation. I was accompanied by my field team of two local assistants, Andrey MacCarthy and Macdonal Gomez; our boat captain, Emanuel Vallejos, from Deep Blue Diving; my two student helpers, Dan Stuart and Veronica Koleff; and my colleague Dr. Nathan Robinson.

It is already midday when we catch a male olive ridley turtle and get him into the boat. After I finish gathering my data, Nathan begins collecting ectobionts—critters living on the turtle—and finds a strange-looking encrustation in one of the turtle's nostrils. We assume it's a barnacle. As I have finished what I need to do and because I think it might be interesting, I pick up my camera to film Nathan as he tries to extract the mysterious object from the turtle's nose.

Andrey and Dan hold the turtle on the bench seat while Nathan carefully pulls on the encrustation using the pliers on his Swiss Army knife. When he's pulled a short length of the object out of the turtle's nostril, we realize it isn't barnacle shaped at all. It looks more like a worm or a worm tube. Through the viewfinder of my camera, I can clearly make out black stripes along its sides. It bears a strong resemblance to a plastic straw. But surely it can't be.

After a few more minutes of gentle pulling, we decide to cut off a piece of the tube-like object so we can take a closer look to see what it is made of. We are curious and want to know what we are dealing with. We also want to make sure we aren't hurting the turtle unnecessarily by pulling it out. Done. The cut-off section does indeed look like plastic. Surely not. Macdonal bravely picks it up and bites down on it. "Yup, that's plastic," he says matter-of-factly. This poor sea turtle really does have a plastic straw stuck up its nose.

There is no way we can release the turtle back into the ocean with a piece of plastic still up its nose. Unfortunately, we don't have the option of taking it to a veterinarian either. So Nathan steels himself and pulls on the remaining section of the straw until it finally comes free and slides out of the turtle's nostril. Nathan holds it up to the camera for me and at exactly that moment the display goes black and the camera stops recording—the battery is dead. At least my hands are now free, and I disinfect the nostril with an iodine-soaked cotton swab to the best of my abilities before we gently hold the turtle over the side of the boat and allow it to slip back into the ocean. Sensing it is free once again, it swims away as fast as its flippers can carry it as we gaze after it, still in shock.

After that, no one is in the mood to collect any more data, and so we start back to the harbor, which is about two hours away. No one speaks on the boat ride back. We are all lost in our own thoughts and when we catch each other's eye, we shrug and shake our heads. We still can't quite believe what we have just seen and done. It seems like a bad dream—except we have the video to prove that it isn't. But how much has the camera captured? The suspense is almost unbearable. As soon as we get back to our quarters in town, I grab the camera's memory card and my laptop. And there it is: over eight minutes of the ordeal recorded with sound and in color. I fast forward to the end and finally allow myself to breathe. In the final frame, Nathan is holding the entire length of the straw up to the camera.

We decide to upload the video to YouTube. After consulting with my academic adviser, I opt to upload the complete video unedited. I don't want to censor or minimize the raw emotions of the moment. I want people who watch it to experience the same shock and guilt we felt. After all, there's hardly anyone who can claim they've never in their life used a plastic straw.

Using the slow Wi-Fi connection from the pizza restaurant next door, I upload the video overnight. The next morning, I don't have much time before I need to be back out on the boat. Hastily, I make the video public. Then I'm up and out of there, spending the next eight hours out on the ocean without any internet connection or cell phone reception. When we return, the video already has a few thousand views. I make a Facebook post, link it to the video, go to bed, and the next day, I'm back out on the boat.

The following evening, I'm stunned. When I open my inbox, I find hundreds of emails from different media outlets requesting interviews and wanting to report on our discovery. None of us expected our video to go viral. The next few weeks are chaotic as we attempt to satisfy the media while having limited access to the internet. We see this as our chance to talk about the problem of plastic in our oceans and the status of sea turtles generally. It is important to us that people understand how much plastic is out there in our environment, the dramatic consequences it can have for marine life, and what we all can do to reduce the amount of plastic, and especially single-use plastic, in our lives. Our message doesn't always hit home. Many media outlets run with the idea that plastic straws are the source of all evil, even though what they really are is a symbol for a world we have fashioned where plastic reigns because it makes our lives so convenient.

Media interest in our discovery is intense but fizzles out just as quickly as it flared up, and we assume it is all over.

People have been momentarily shocked, but now everything will return to normal. However, a few months later, I begin getting inquiries from organizations working on the issue of plastic pollution, and less than a year after the video, the first campaigns pop up that try to persuade people to use fewer straws and less single-use plastic in general or, preferably, to not use plastic at all. The ripples made by my video are being amplified into ever-larger waves, and the world is gradually waking up to the issue of plastic pollution in our oceans. Powerful images in films such as *A Plastic Ocean*, which was released in 2016, and programs such as David Attenborough's *Blue Planet* II bring the message to an even wider audience.

Much has happened over the past few years. Various countries have passed laws banning single-use plastics, and awareness of the whole issue has grown exponentially. People all over the world are trying to reduce their plastic footprint and manufacturers are feeling the zeitgeist. Hundreds of plastic-free stores are springing up, and the plastic-free packaging of previous generations is making a comeback. Even so, there is much that still needs to be done. Legislation needs to force the largest plastic polluters, companies such as Coca-Cola, Nestlé, and the like, to take more responsibility. But even they are realizing that their customers don't want plastic anymore and are searching for alternatives.

Unfortunately, right now plastic is our only choice for a cheap material with such a wide variety of applications. The hope is that a similarly versatile, mass-produced, but biodegradable material will be discovered that will render plastic obsolete. I am highly optimistic. If our big brains can fly us to Mars, they should be able to solve this problem too. Sometimes, of course, you need the right motivation. Money always helps here. Calculations suggest that switching to a closed-loop or circular economy in which all materials are completely reused, recycled, or composted could reduce the amount of plastic that ends up in the oceans by over 80 percent by 2040 and reduce the amount of new plastic that is manufactured by 55 percent. That, in turn, could lead to savings of up to US$70 billion for governments worldwide by 2040. A side benefit is that greenhouse gas emissions would be reduced by 25 percent and an additional 700,000 jobs would be generated, mostly in the Global South. In 2022, heads of state, ministers of the environment, and other representatives of United Nations member states agreed on a historic resolution to end plastic pollution and to forge an international legally binding treaty by 2024.

This development gives me hope, and I finally feel my work might make a difference after all. Even my PhD research helped. The data I collected contribute to our knowledge about sea turtle migrations, and I hope they will help us create better and more effective means of protecting turtles. I fitted a total of twenty-seven female olive ridley turtles with transmitters, which allowed me to follow them on their journeys after the nesting season. Although olive ridleys are nomadic, my data helped to identify specific places along the Pacific coast of Central America where large numbers of them can be found. These regions should be at the top of the list for Marine Protected Areas dedicated to the conservation of the Eastern Pacific population because this would be one of the best ways to protect them.

We still don't know enough, however, about sea turtle migrations, and it is extremely important to learn more about where they spend extended amounts of time. Only when we identify specific areas can we provide truly comprehensive protection for their nursery areas, feeding grounds, and migration corridors. This is why I have been focusing my research on exactly this topic for the past few years, and I dare to hope that my data will help ensure the existence of sea turtles in years to come.

KEMP'S RIDLEY TURTLE

CR

Lepidochelys kempii

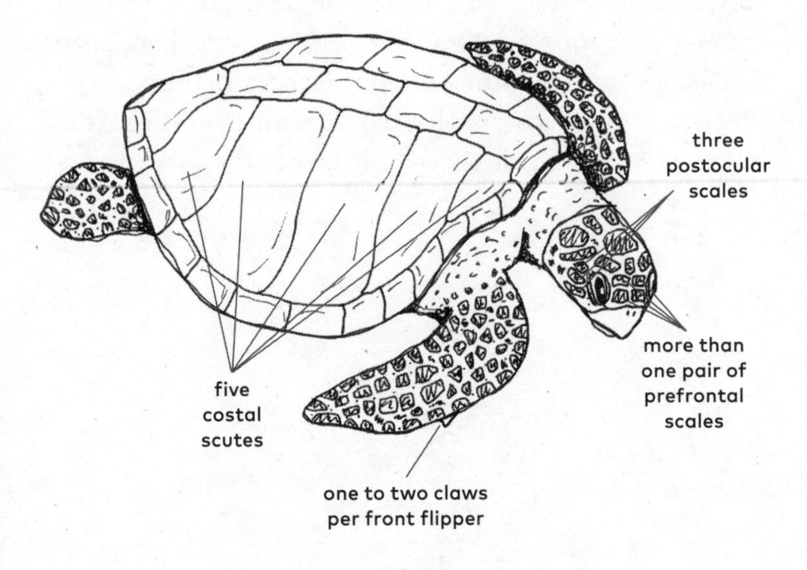

IUCN STATUS: CR—Critically Endangered

CARAPACE LENGTH: 55–70 cm (22–28 in.)

WEIGHT: 35–45 kg (77–99 lb.)

MAIN DIET: Crustaceans, especially blue crabs

AGE AT MATURITY: 11–16 years

REMIGRATION INTERVAL: 1 year

INTER-NESTING INTERVAL: 20–28 days

NUMBER OF CLUTCHES PER NESTING SEASON: 2–3

NUMBER OF EGGS PER CLUTCH: 105

INCUBATION TIME: 45–60 days

GOOD TO KNOW: Range mostly restricted to the Gulf of Mexico; smallest species of sea turtles; synchronized mass nesting (arribadas) and the only species to nest during daylight

– 6 –

SEA TURTLE LOVE

WE ARE DRIFTING in a small fishing boat in the calm, deep-blue waters of the Pacific Ocean a few miles off the west coast of Costa Rica. Two olive ridley turtles are sharing the shade of the boat's awning with us. One is lying on its back in a car tire, and the other is being held on the bench seat by two of my assistants. Both have a dark, damp cloth over their eyes to alleviate their stress. One is male and the other is female, and we've just watched them in the act of mating. As soon as they finished, we snorkeled over and fished the two of them out of the water by hand. They are part of a genetic study that we hope will shed light on the mating preferences of sea turtles. We take skin samples from both turtles and meticulously document morphological differences between them. Weight, length and width of the carapace and plastron, length of the tail, length of the claws, width of the head, length of the flippers—all measured, recorded, and photographed.

Because sea turtles mate in the water and not on land, we still do not know much about what goes on. What we do know is that the male and female must meet and copulate. That might not sound like groundbreaking information, but not all animals need to have sex to produce offspring. For many animals other than mammals, birds, or reptiles, external fertilization of the female eggs by the male sperm is the norm. With sea turtles, however, fertilization takes place inside the body, just as it does for humans. And for that to happen, the male and the female must first find each other.

That might not sound too difficult, but sea turtles are not especially social. While the film *Finding Nemo* gives the impression that sea turtles live in family groups with a patriarchal structure, the reality is that each turtle does its own thing after leaving the nest along with its siblings. By the time the babies hatch, the mother is long gone and the father was never in the picture anyway. Each turtle fends for itself. Males and females don't meet up again until they are sexually mature and it's time to mate. Somewhat surprisingly, mating takes place not in the turtles' feeding grounds but mostly hundreds or thousands of miles away in coastal waters directly in front of the females' nesting beaches.

We don't yet know exactly what happens during a turtle's very first journey in search of a mate or what triggers their migrations. In general, it seems that males and females reach sexual maturity when their carapace length is slightly less than average for the adult population, and this is when they mate for the first time.

The reproductive cycle in both males and females is controlled by hormones, including the male sex hormone testosterone. Courting and mating behaviors have been observed in experiments in which immature sea turtles were injected with testosterone. In young sea turtles, the secretion of testosterone appears to be temperature dependent, but scientists believe that once puberty hits, the release of this hormone is controlled by light. Sea turtles have a well-developed pineal gland. This is a small endocrine gland in the brain of most vertebrates. It secretes melatonin, a hormone that is produced and released into the blood when it becomes dark outside, so mostly at night. Melatonin influences sleep-wake cycles and other time-dependent rhythms of the body, including the reproductive cycle. A malfunction in the secretion of melatonin can lead to disrupted diurnal rhythms, and also to premature or delayed sexual maturation—including the late development of sexual characteristics. From the perspective of evolutionary biology, the pineal gland is a kind of atrophied photoreceptor. In the brains of some species of amphibians and reptiles, it is paired with a light-sensitive organ called the parietal or pineal eye, which cannot make out shapes but can distinguish between light and dark. Mammals, birds, and many reptiles lost the pineal eye over the course of evolution, but the pineal gland remains and still responds to light, which is why the pineal gland is sometimes called the third eye.

At some point in a sea turtle's development, the pineal gland probably assumes control of testosterone secretions. This would be the point at which light intensity rather than temperature regulates the onset of puberty and the turtle's journey to eventual sexual maturity. It goes without saying that to reach this stage in life, a turtle also has to be able to find enough food and all its organs need to be fully developed. As the feeding grounds of most sea turtles lie at higher latitudes than their tropical nesting areas, and as day and night length vary considerably during the annual cycle in these regions, it could be that the pineal gland monitors day length over the course of a year and triggers the turtle's first migration—as well as all subsequent migrations—to find a mate, allowing females and males to arrive in their mating areas at precisely the right time to meet up with others of their species.

Before these long mating migrations begin, however, males and females need to fatten up in their feeding grounds. The stored fat in males fuels production of sperm, and in females fuels the production of mature ovarian follicles. Each follicle in the female's ovary consists of an egg cell surrounded by auxiliary cells. A precursor of egg yolk known as vitellogenin helps transport protein and some lipids from the liver through the blood to the growing egg cells to form nutrient-rich egg yolk. The yolk, which provides the building blocks for the embryo's growth, is made up of protein and fat in a ratio of 1.6 to 1. The more abundant the food sources, the fatter the male and female turtles get and the more likely it is that the females especially will set off the following season to lay eggs on their nesting beaches. Because it takes a lot of energy to produce egg follicles and swim long distances, most females embark on this arduous voyage only once every two to three years. When food is scarce, the gaps between journeys can be much longer. The *remigration interval* is the technical term scientists use to describe these gaps between the females' journeys. Males, whose sperm production is less energy consuming, usually make the journey every year.

For a long time nobody knew how females and males choose their breeding areas. Scientists did realize early on that females remain true to their nesting beaches for decades, but the reason was unknown. It was speculated that the first time young adults embark on these journeys they follow experienced sea turtles to their breeding and nesting areas, imprint on these locations if they too meet with success, and return to them in subsequent seasons. The question got a definitive answer only after someone had the inspired idea to put so-called living tags on hatchlings.

The metal tags we use for adult turtles are much too large for babies and too expensive. Because so few small turtles survive to adulthood, you would have to tag an enormous number of them to find just one a few decades later. Living tags, however, cost almost nothing. This technique exploits the distinct difference in coloration between the dark upper shell (carapace) and the light lower shell (plastron). This color difference camouflages swimming turtles: predators below turtles have difficulty distinguishing them from the bright surface of the water, and predators above turtles have difficulty distinguishing them from the dark ocean floor. For living tags, small circular pieces of keratin are cut from the upper and lower shells and transplanted to the opposite shell. Imagine us cutting pieces from our fingernails and integrating them into our toenails. These tiny but eye-catching circles fuse with the keratin around them and increase in size as the turtle grows. For the rest of the turtles' lives, these circles and their position on the carapace identify them as members of a particular study group. The small light circles on the dark carapaces of nesting sea turtles are clearly recognizable even in the dark of night. Living tag studies revealed that it can take turtles from fifteen to forty-five years to reach sexual maturity and established that adult females return to the beaches where they hatched, or to beaches close by, to lay their eggs. Genetic studies have since confirmed these findings and shown that males also travel to breeding areas near the beaches where they hatched.

I too had to travel far to find my life partner. Unlike the sea turtles, however, I did not find him in the place of my birth but on the other side of the world. At some point, my colleague and good friend, Andrey, who explained to me right at the beginning how to dig a sea turtle nest and helped me see the beach and the sea turtles as the local people view them, became more than a friend. We've been a couple for many years now and he often shares my many sea turtle adventures.

Now, on the boat, we are working as a well-coordinated team. After just twenty minutes, we release both sea turtles back into the water. We spot three more olive ridley turtles swimming close to the surface not far away. Two of the turtles appear to be taking a great interest in the third. I put on my dive mask, pull on my flippers, and slip over the side of the boat and into the water as unobtrusively as possible. One of my assistants hands me my GoPro camera and I dive. I carefully swim toward the turtles. Luckily, I find myself directly behind them; they haven't noticed me yet. Under the water, I can see that, as I already suspected, what we have here is one female and two males. The massive, overly long tails of the males are hard to miss. Both are swimming in circles around the female and trying to climb onto her back. The female doesn't let them out of her sight and keeps her head and chest turned toward the males. She also has her hind flippers pressed up against her tail, a clear sign she is not in the mood to mate. But that doesn't stop the males from trying. Round and round they swim, biting the female's neck and flippers as though that might change her mind. Noticing a moment of inattention, one of the males makes a daring dash toward the female and attempts to pull himself up onto her back. She turns quickly and the determined suitor falls off. Clearly neither of the males is familiar with the concept of consent. At some point it all becomes too much for the female, who probably only came up to the surface to warm up a bit. Taking advantage of a gap between the circling love-crazed males, she races off into the unending blue with powerful strokes of her flippers.

I wouldn't call the mating behavior of sea turtles in any way romantic from the human perspective. Once a male manages to pull himself onto a female's back and gets a firm hold by grabbing the edge of her carapace with the claws on his front and hind flippers, he's difficult to dislodge. He has a lot of control over his long prehensile tail. He can use it to grab the female's tail and feel for her cloaca so he can insert his penis. The probing tails and penises of males lead to many amusing scenes on the boat: if you're not paying attention, you might suddenly feel a sea turtle penis tapping on your hand.

Cloaca is the Latin word for the sewers ancient Romans built to drain wastewater from their cities. In many animals, such as birds, reptiles, and amphibians, the cloaca is a common opening in the body used by digestive, sex, and other excretory organs. That's how it is with sea turtles too. In addition to urine and feces, sex-specific products also exit the body via this opening: eggs from the female and ejaculate containing sperm from the male. The penis is also tucked away in the cloaca when it is not engorged.

A sea turtle's penis is a curious organ, a kind of hydraulic half cylinder well supplied with blood vessels. When a male turtle is aroused, this singular erectile body engorges with blood. This makes the penis relatively stiff, which it needs to be if the ejaculate is to travel through it because a sea turtle penis does not have an inner tube or duct. Instead, when a sea turtle penis engorges, it forms a long groove through which the semen travels. The raised areas on either side of this groove are known as seminal ridges. The groove stretches from the urethral opening at the base of the penis to the glans at the tip. It becomes a conduit for sperm only when the penis is engorged.

That might seem unusual to us, but in fact the sperm duct in a mammalian penis is the exception. When you look at the male sex organs of lizards, snakes, crocodilians, and birds, grooves and seminal ridges are the norm. When a sea turtle penis engorges, its length increases by almost 50 percent and its diameter by 75 percent. Even a flaccid penis is rather large. Flexible furrows in the skin and a kind of hook at the tip make it look even more strange to us. Turtles seem to be able to control the movements of the seminal groove and the furrows at the tip of the glans, which presumably helps them direct their ejaculate to the right spot within the female.

The female's sex organs are a little less alien looking. From the outside, only the cloaca is visible. Females have a pair of ovaries and a pair of oviducts inside their bodies, just like female mammals do. Unfertilized egg cells start to ripen in the ovaries about eight months before it's time to mate. Sperm joins ripe egg cells with their nutrient-rich yolks in the ovaries. Both are then enclosed in a shell so the female can lay the fertilized eggs.

After mating occurs, the female stores the sperm and uses it throughout the season to fertilize the multiple clutches of eggs she will lay. It takes a female from ten to fourteen days to prepare a complete clutch of eggs, and this interval determines the female's egg-laying cycle. These inter-nesting intervals vary depending on the species. Leatherback turtles, for example, nest every ten days, whereas hawksbill turtles nest every fourteen days. Olive ridley turtles can delay laying their fertilized eggs even longer and lay every seventeen to twenty-eight days.

On the boat, we are particularly interested in the female's reproductive status. In other words, we want to determine at what point she is in her cycle because we don't yet know for certain whether females mate only once at the very beginning of the nesting season or if they might mate again later between nesting events. When I pull our next female out of the water, I use our portable ultrasound machine with its probe to look at her oviducts and ovaries through the inguinal region, the soft groin tissue between her hind flippers and plastron. Carefully, I set to work.

Her ovaries still contain yolk-filled follicles—egg cells that have not yet ovulated. They look like hundreds of tiny Ping-Pong balls. This could be a sign the female has just arrived in the breeding area. Or maybe she just laid a clutch of eggs last night. Whichever it is, she will be laying another clutch of eggs. Eggs enclosed in a shell look like larger follicles but with a kind of halo around them. As we continue the ultrasound, we tip the female up a bit to allow gravity to shift her internal organs closer to the probe so I can get a better view of her oviducts and ovaries. Even so, it's difficult to get a clear ultrasound image because the female is squirming and writhing with all her might and we're working up a sweat trying to keep her body and flippers still.

But back to our original question. To discover when the female mated and with whom, we first need to understand the criteria she uses to choose her partner. There is no open combat between males, and females don't choose the most beautiful, largest, most colorful, or cleverest partner—at least not as far as we know. It would probably be challenging to choose based on external characteristics because apart from the long tail and massive claws on the males, there are not many attributes that set the sexes apart. Instead, mate choice seems to follow what is known as a scramble mating system, in which mating success is dictated by a speedy search for a partner. Basically, it's first come, first served.

In a system like this, a turtle could end up mating with another turtle that might not be, genetically speaking, the fittest. Or the pair could be genetically incompatible and produce offspring with poor chances of survival. For males, that's not a great loss. They try to pass along as many of their genes as possible to the next generation and mate with as many different females as possible in a single season. For the females, however, a poor match means that they will have invested all their energy and resources for nothing if none of their babies survive that season. Therefore the females also mate with multiple males, probably hedging their bets by improving the chances for their offspring with a random partner upgrade.

And so there is a classic conflict of interest between the sexes. Females need enough sperm to fertilize all their maturing eggs, but they also want it to be of the highest possible quality to ensure the best possible genes for their babies. Males try to father as many offspring as possible with as many females as possible. This explains why males remain attached to their partners even after copulation is over, to prevent other males from mating with them. They sometimes remain attached for hours. As long as a male is on top of her, the female has to drag twice her own weight through the water, which must be especially difficult when she has to come up to the surface to breathe. Other males also often try to pull their successful competitors off the female's back or even pile on top of her. A female can end up with five to seven males weighing her down, with none of them paying any attention to her. The mating season starts one to two months before the onset of the nesting season, and in larger populations you see hundreds of exhausted females bobbing listlessly at the surface with males hanging off them.

Of course, we shouldn't anthropomorphize animal behavior. However, sometimes when I see females with bite marks on their necks, and gashes and scars on the soft tissue along their shell—injuries inflicted by the beaks and claws of males—I can't help but wonder what they think of it all.

A somewhat humorous aspect of sea turtle love is that the males are not particularly picky about the objects of their desire. There are, for instance, thousands of reports of males trying to copulate with floating driftwood. And some divers have described having to fight off the aggressive approaches of an amorous male. Perhaps that explains why hard-shelled turtles often interbreed, leading to crosses known as hybrids, which, like mules, inherit characteristics of both parents. But, unlike mules and similar crosses, these hybrid turtles are capable of reproducing, which suggests that the genetic differences between species do not run as deep as those between horses and donkeys.

A PREVIOUSLY SUSPECTED CROSSBREEDING also lies behind the unusual name of the two ridley turtles, which are known as "bastard turtles" in German. The English name *ridley* probably comes from *riddle* and points to the puzzle these two species posed to science for a long time. The story of Kemp's ridley turtles began when Key West businessman and avid naturalist Richard Kemp discovered a sea turtle that the locals called ridley. Kemp was fascinated by this strange turtle, which looked so different from the ones he knew from the Bahamas. He sent descriptions and specimens to Harvard, and in 1880 the species was named Kemp's ridley in his honor.

But Kemp's ridley turtles were still confounding scientists into the twentieth century because no one knew where they nested. For this reason, the father of sea turtle biology, Archie Carr, traveled to the Florida Keys in 1938 to listen to local stories about this strange-looking turtle. But even the fishers did not know where they nested; they thought the animals were hybrids of green and loggerhead turtles, so-called bastards.

Carr, who unlike the fishers knew that Kemp's ridleys were a distinct species, spent the next two decades searching all around Florida, the Bahamas, the Caribbean, and the northern Gulf of Mexico for their nesting beaches. He never found them. Then, in 1947, a pilot filmed amazing scenes of synchronized mass nesting of Kemp's ridleys on a relatively short section of beach in Rancho Nuevo, Tamaulipas, Mexico, during which an estimated forty thousand turtles dug nests all at the same time—an arribada. But scientists didn't hear about this until 1961 when the biologist Henry Hildebrand stumbled upon the film on a visit to Mexico and presented it to the scientific community. The two biggest riddles about the Kemp's ridley turtle were now solved: it was definitely not a hybrid but a species in its own right that reproduced selectively with others of its own kind, and the females nested in arribadas in Mexico. The two species of ridleys—Kemp's and olive—are the only species of sea turtles in which large numbers of individuals synchronize to nest en masse.

There are many different stories about both species of ridley turtles, which used to be thought of as evil. They were known as "heartbreak turtles" because they fought so vigorously when they were turned over onto their backs that, it was said, their hearts exploded and they died. We have regularly experienced their pugnacity on our boat—thankfully without any exploding hearts. As it turned out, a ridley turtle provided me with my most physically painful turtle interaction to date.

It happens during one of our in-water surveys. I am totally exhausted as I pull myself over the railing and into the boat after our first successful catch of the day. Unfortunately, I land right in front of the beak of a female olive ridley lying on her back in our car tire with a cloth over her eyes, waiting to be examined. My assistants must have been asleep on the job because our rule is to turn the turtles with their heads pointed to the center of the boat and away from the railing so I can easily climb in and out. Before my brain is even aware of the danger, the female bites into my exposed shin. I feel a sudden searing pain as a good chunk of my leg is caught in the sea turtle's mouth. She is not about to loosen her grip one iota. Now, just don't try to pull your leg away, I tell myself as I grit my teeth to help get me through the pain. If you do that, you can say adios to that piece of your leg she has in her mouth.

Looking up, I see the ashen faces of my assistants. Captain Santiago is rummaging feverishly in the back of the boat for an object he can use to pry the turtle's mouth open. I experience it all in slow motion and the pain is almost unbearable. The last thing I want, however, is to spread panic, even though in my mind's eye I can already see myself in the hospital being stitched up. At least it happened to me and not to one of my assistants, I think. After a few seconds—which seem to me like an eternity—the female opens her parrot-like beak a fraction of an inch and I react at lightning speed. I yank my leg back as fast as I can. I am mentally preparing myself for a bloody mess, but under the circumstances I was lucky. The female bit me through a beakful of cloth. Her jaws clamped down on my skin with tremendous force but she didn't actually tear away any flesh.

Still, it isn't a pretty sight. Despite the cloth, the two sharp tips on her beak have left bloody marks, crushed skin, and a bruise that will bloom into a dazzling array of colors over the next few days. My usually laid-back assistant Marcus has eyes as big as saucers and anxiously asks if I'm okay. As I don't like being the center of attention and I'm also worried about the well-being of the two turtles we have on board, I act as though there is nothing to see and we should just get on with things. It is only when the two turtles are back in the water that I allow myself to take a deep breath, sit down, and eat a granola bar. I notice that with all the shock and stress, my blood sugar level has plummeted and I am trembling uncontrollably. My leg really hurts, and the spot where the turtle bit me is throbbing. We are all deeply shaken, but one good thing comes out of this experience. From then on we are twice as careful when we position turtles in the boat, as this incident has been most instructive for all of us.

WHILE I SIT on the bench munching my granola bar, I can't help but ask myself what in the world I am doing here. It all started innocently enough with my first trip to Costa Rica when I was a master's student and neither the locals nor a later intensive search of the scientific literature could answer the many questions I had about the mating behavior of sea turtles. And so, when I was offered the opportunity to do my master's project in Costa Rica, it didn't take me long to settle on a topic. I had no doubts. I wanted to research the mating behavior of leatherback turtles, focusing on how often males and females mated each season and with how many different partners. Up until a few years ago, these turtles were thought to be promiscuous—because of their scramble mating system—but nobody knew whether that was the case for both sexes. Back at the University of Würzburg, Dr. Heike Feldhaar and her research team took me under their wing and supported my work. Dr. Feldhaar is a well-known ant researcher who uses molecular tests, in other words genetic testing, to investigate family relationships and paternity in ant colonies. The same methodology can be applied to other organisms that reproduce sexually, and so I decided to do a genetic paternity study of the leatherback turtle population in Gandoca.

The question I wanted to answer with my master's work was exactly the kind of puzzle I have loved ever since I was a child—the difference being that back then I found the answers in books. This time I would have to find the answers myself by carefully collecting data and samples, analyzing these samples in thc laboratory, and evaluating my results. Armed with my main tool, genetic analysis, I set out to uncover the family relationships of baby leatherbacks in Costa Rica and fill the gaps in our knowledge about the mating systems of this species. I was especially interested in whether individuals of both sexes in one population generally mate with different partners (polygamy), or whether only females mate with many different males (polyandry), or whether only males mate with many different females (polygyny). Using genetic testing, I hoped to detect how many different fathers sired the babies in one female's nest. This would also allow me to determine how many males she had mated with.

The principle behind paternity analysis is relatively simple. Anyone who remembers biology lessons in high school, and the name Gregor Mendel in particular, might also remember that humans and all other animals that reproduce sexually inherit 50 percent of their DNA from the mother and 50 percent from the father. And so, we usually have two variants, or alleles, of each gene that codes our physical characteristics. Our genetic makeup—in other words all our genetic information taken together—is called our genotype. For instance, we could have one allele for blue eyes and one allele for brown eyes. Certain laws dictate, however, which of the two alleles ultimately determines our actual eye color and thus our phenotype (the physical expression of the gene).

To analyze paternity, however, all we require is the genotype of the mother and her children, for which it is usually sufficient to determine the genotype of a handful of microsatellites. Microsatellites are repetitive, noncoding sequences of DNA that are described and identified by their physical position—their locus—in the genome. So, if I look at the mother's two alleles at the loci I have chosen and compare them with the two alleles of her offspring, I can see which allele has been inherited from the mother, and which, in accordance with Mendel, must come from the father. Of course, the males also have only two alleles per locus and if I find more than two paternal alleles in a nest, I know that more than one male sired the clutch I am testing, which makes some of the baby turtles half siblings.

Yet it's difficult to come to definitive conclusions because some questions remain unanswered, including how long females can store sperm. We know they are normally capable of doing so at least for the many months of the nesting season, because females don't lay just one clutch of eggs per season but, depending on the species, from an average of two, as with the olive ridleys, to as many as seven, as with the leatherbacks. There are, however, also stories about females in aquariums still laying fertile eggs after years without any contact with a male. If these stories are true, it makes it difficult to ascertain whether the second or third fathers of a clutch bred with the female in the current season or in a previous season. Long-term sperm storage does, however, raise the question of how long sperm remains viable and whether its quality declines over time and the genetic material it contains gets damaged. For these reasons, many scientists surmise that the sperm used to fertilize the eggs usually comes from the most recent mating, presumably following the maxim "last in, first out."

For my paternity research, I have to take skin samples from lots of females and their offspring, preferably from more than one nest. That sounds easy, and the first skin sample I take from a female certainly is not a big challenge. I wait on the beach at night until my huge moms are finished laying their eggs. Then I use a vascular clamp to pinch off a small section of skin on the edge of one of the hind flippers and snip off a 6-millimeter (quarter-inch) sample using a sharp scalpel blade. Without allowing the sample to touch the sand, I drop it into one of my prepared and numbered vials, which all contain 99 percent ethanol. Most of the females I take samples from don't even flinch. It is the same thing later with the babies. When the eggs hatch in one of the hatchery nests, I pack ten babies into a cooler and carry them back to camp, where I have a table and better light. The process is the same as with the mothers except the skin samples are infinitely smaller—so small that they are almost invisible in the alcohol in the test tubes.

The problem is how to keep track of all the females over the course of the nesting season and safeguard all their subsequent nests as the season progresses. To do this, I tape lists with my females' tag numbers to the clipboards in the patrol backpacks and instruct my colleagues to make sure they take each clutch of eggs from these females to the hatchery. Then I have to cross my fingers for the next sixty days and hope that the eggs will hatch without any problems and no inattentive volunteer will release the babies to the ocean before I have a chance to take skin samples.

Naturally, it's boring when everything goes as planned. Because life loves to throw curveballs, the community of Gandoca gets into a dispute about who should be in charge of the sea turtle project, and the government ends up dividing the beach between the two opposing factions in the middle of peak nesting season. A decision reminiscent of Solomon's, perhaps. Unfortunately, one of the two hatcheries with my clutches of eggs inside is now on the section of beach my group is no longer allowed to patrol. We have about eighty nests in that hatchery, including about thirty that are part of my master's research. I have to resign myself to the fact that these data are now lost to me and my study. But then after a month of tough negotiations, we are allowed to at least continue monitoring the hatchery and release the babies after they hatch, even though we are not allowed to relocate new clutches of eggs inside. Still, this is a definite improvement.

For the next weeks, things go smoothly under this new arrangement. But then in June heavy storms and monsoon-type rains push high waves onto the beach. We erect a wall of sandbags to protect the hatchery from flooding, but the first waves whipped up by this storm are so high they wash all the sandbags away. Like it or not, we must battle the water ourselves. For many days, we try everything to save our hatchery with the seventy nests that remain. Early every morning, we fill new bags with sand and pile them up in front of the hatchery. Then we look for the largest pieces of driftwood and downed trees and build a barrier in front of the sandbags. But every night our fortifications disappear within minutes when the tide comes in.

In a final desperate attempt, we gather all our assistants and volunteers to move our nests to a different location. We divide everyone into three groups. The first group fills boxes with a layer of sand and while the rain pours down on us, we dig up each and every remaining nest and lay the eggs inside the boxes as carefully as we can in the dark and chaos, making sure not to turn them upside down. It is of utmost importance to make sure we label the boxes correctly so we don't lose track of the eggs and the identity of their mothers. The second group carries the boxes about half a mile to another section of beach that is not impacted by the waves. There, a third group of assistants is busy digging nests in a new, and hopefully safe, spot. They dig the nests one after the other as though they are working on an assembly line. Tonight they are breaking production records.

I am worried about the eggs, and so at the beginning I oversee the first group to make sure all the clutches of eggs are packed and identified correctly. As the evening progresses and routines gradually become established, I hurry to the third group to offer assistance as the nests are dug and to make sure the eggs are safely reburied. I'm soaked to the skin, my hair is sticking to my face, and I'm covered in sand from head to toe. The rain doesn't let up, and as I carefully transfer the eggs from the boxes into the egg chambers that have been prepared for them, I try with little success to shield them from the worst of the rain with my body. My muscles are screaming, my back is hurting, and I emptied my water bottle long ago. I know I'm not the only one who is exhausted and miserable, but we continue to work furiously through the night and when the sun rises at five, we have done it. The last eggs have been safely reburied.

A dismal scene greets us at first light of day. The strip of beach where we have moved the eggs looks like a minefield with a dark patch where each new nest has been dug. When we head back to our former hatchery, we meet our comrades in arms from the two other groups who, like us, are a miserable sight: exhausted, drenched, covered in black sand, and shivering from the cold. The hatchery itself, or what is left of it, looks as though a tsunami has washed over it. The fence is half torn away, the camouflage net on bamboo poles that provided shade has collapsed, and the hut for our equipment is no longer standing. We all have to take a moment to let the events of the previous night sink in. Feelings of deep despair about the condition of our hatchery alternate with feelings of euphoria that we have managed to save all the nests. All the tiny turtles have to do now is hatch. But we can worry about that later. First a cold shower and then to bed. This is one of the rare days when I wish I could take a warm shower instead of a cold one.

Luckily, all our nests hatch with about the same level of success as those in an undisturbed hatchery, and I have no difficulty collecting the samples I need for my genetic study. I am surprised when the results of my research show that only a small number of females have mated with more than one male, probably because leatherbacks are so endangered. When you compare my results with other species and populations, it is clear that the larger the population, the higher the percentage of females mating with multiple partners, likely because in populations with more individuals, there are also more sexually mature males available and therefore they are easier for the females to find. It is concerning to discover that population density is limiting the ability of sea turtles, and my studied leatherback population in particular, to mate with multiple partners. It is just one of many signs that this population is too small. At this rate, it will not survive. It needs more individual turtles. Genetic diversity is critical for the future of any population because it provides the building blocks a species needs to adapt to its environment.

As I write my thesis, I learn more and more about the mating systems and behaviors of sea turtles, but new questions constantly arise. What criteria do males use to choose females and females use to choose males? How do turtles that want to mate find each other in the vastness of the ocean? Is there some kind of a meeting spot or are they guided by scent? Can sea turtles recognize relatives? If not, how do they avoid mating with their parents, siblings, aunts, uncles, or grandparents, all of whom return to the area and even to the individual beach where they got their start in life?

In the years that follow, I turn my attention to other aspects of sea turtle conservation and research, but I cannot stop thinking about these questions. And so it makes sense that they are first and foremost in my mind when I finally decide to embark on a PhD. I lay out my questions when I apply to four different research groups at universities in the U.S., and they lead me to my eventual PhD adviser, Dr. Pamela Plotkin in the Department of Oceanography at

Texas A&M University, who shares my research interests. And that is how I find myself sitting in a small fishing boat out on the Pacific Ocean researching mating sea turtles. Unfortunately, after the first season, I have to radically revise my list of questions because my doctoral committee fears they cannot all be answered in just one dissertation. They are probably right...

When I look back at the path my life has taken, I can't help but be deeply grateful for sea turtles and my love of them—not only because of all the knowledge I have gained and the exciting questions I get to answer, but also because I am keenly aware that my desire for adventure and search for the next thrill could easily have led me in a very different direction. My love for the ocean and later my love of sea turtles became my North Star very early on and they always guided the next steps I took. These animals and their protection are the reasons I get out of bed in the morning and why I can never give up. Protecting sea turtles is what I was put on this earth to do. They are my purpose and my destiny. They are what I live for.

I think having a sense of purpose explains why, even at a young age, my vision for my future was so different from those of my classmates. I wanted to save the world and the whales and have adventures. Their plans revolved around fat paychecks, home ownership, and cars. None of that has ever meant much to me. Of course, having a regular income makes for an easier life, but in the end having more money doesn't guarantee happiness. Quite the opposite. Over the years, time and again people find their way to my projects who, even though they have had successful careers by society's standards, have arrived at a point where they are questioning their purpose in life. Our sea turtle projects have given more than one person a new outlook on life and have often led them to abruptly reevaluate their priorities.

That makes me value my guiding star even more today. It makes it easier for me to engage in difficult situations, weather setbacks, and still never lose hope and motivation. I can only underscore how important it is for everyone to follow their passions and find their purpose. "Find a job you enjoy doing, and you will never have to work a day in your life." Of course, you can only afford that much introspection when you have the privilege of choosing how to lead your life. Not everyone has this luxury. While I am floating weightless in the deep blue of the ocean watching two sea turtles mate, I can't help but be deeply happy. The water is crystal clear and I feel like I'm flying. There is nowhere I would rather be, and I am infinitely grateful.

FLATBACK TURTLE

Natator depressus

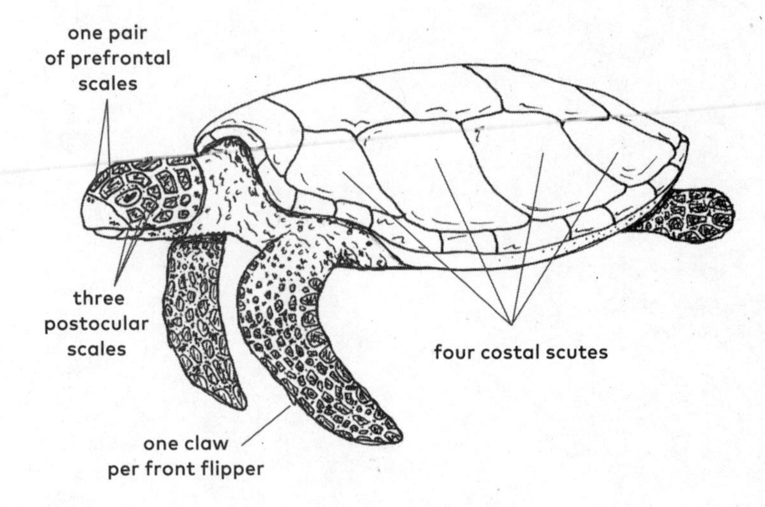

IUCN STATUS: DD—Data Deficient

CARAPACE LENGTH: 75–99 cm (30–39 in.)

WEIGHT: 70–90 kg (154–198 lb.)

MAIN DIET: Soft invertebrates

AGE AT MATURITY: 15–25 years

REMIGRATION INTERVAL: 2–3 years

INTER-NESTING INTERVAL: 15 days

NUMBER OF CLUTCHES PER NESTING SEASON: 3–4

NUMBER OF EGGS PER CLUTCH: 55

INCUBATION TIME: 50–55 days

GOOD TO KNOW: Range is limited to coastal waters of northern Australia, southern Indonesia, and southern Papua New Guinea; eggs are large in comparison to body size; flat shell, hence the common name

– 7 –

HOMECOMING

WATCHING A SEA TURTLE lay her eggs is probably one of the most moving experiences you can have in nature. Barely any other wild animal allows us to come so close for so long. The first time I saw a leatherback nest was a moment that completely changed my life. When a female materializes in the waves at night like a shadow, makes landfall, and laboriously crawls up the beach, you know nothing is more important to her than this. You can feel the unbelievable power of nature driving her as she struggles to drag her massive body, which only seconds before was floating weightlessly in the water, up 20 to 30 meters (65–100 feet) of beach. She has just one goal: to lay her eggs in the warm tropical sand.

Egg laying for sea turtles is usually a play in seven acts, and the script varies little between species and individuals. It's like a computer program that, once launched, never fails to run through previously established levels of behavior and sequences of movement in a predetermined order. As soon as the turtle reaches a safe place and finds her rhythm, she is basically unstoppable. She could be missing one or more flippers or dragging a fishing net behind her; nothing, it seems, can deter her from completing her mission.

One balmy night, I am patrolling the beach with Andrey and my student Kim looking for olive ridley turtles. During the five-to-six-month nesting season, our team patrols the beach every single night and each team member has one day and one night off per week. Unfortunately, we can't catch up on sleep during the day; not only do we have too many things to do—administrative tasks, beach cleanup, and nest inventories—but also it's too hot in the un-air-conditioned buildings. During my first years in Costa Rica, I even worked two nesting seasons a year: the leatherback nesting season in the Caribbean, followed immediately by the leatherback nesting season in the Pacific. Those were hard times both physically and mentally. Too little sleep is extremely unhealthy for the human psyche; there's a reason sleep deprivation is used as a method of torture. During a nesting season, we are more like zombies than people. Back then, because I was responsible for both my team and the turtles, I often had no more than one or two hours of sleep a night for a couple of weeks, and I turned into a complete grouch. At peak nesting season, I was frequently so sleep deprived that I would just fall asleep wherever I stood or sat in moments of calm. My team soon learned if they woke me for anything other than an emergency, I would fly into a rage and become extremely unpleasant. Luckily, they understood, because lack of sleep is hard on everyone, and during nesting season none of us are particularly nice people all the time.

And so, here I am on a dark night, keeping an eye out for shadows in the water. Even from a distance, I can recognize each species based on how they move. Leatherbacks and green turtles tend to pull themselves laboriously up the beach with both front flippers at once; other species move with relative ease alternating flippers as they make their way toward the vegetation line. As a female crawls up the beach, she frequently stops and tests the air for any lurking predators. Every unusual movement or unexpected light could be interpreted as a threat and cause her to quickly retreat into the water. We assume a female turtle uses her sense of smell to identify the beach where she hatched and to test the chemical makeup of the sand and thus find the ideal spot to lay her eggs.

As I walk along the water's edge, I'm also looking for the telltale tracks a sea turtle leaves as she makes her way up the beach. These are often the first sign of a female's presence. And there it is. By the light of the moon, I suddenly make out a black line that leads directly from the water to the trees. The sight of a sea turtle always releases a rush of adrenaline inside me. No matter how tired I am the moment before, it always wakes me right up, washing away any exhaustion. I squint and look at the line out of the corner of my eye to see it better in the darkness. Only one set of tracks means the female is still on the beach; two means we've missed her. We are in luck. There's just one line and as I follow it with my eyes, I can make out the hunchbacked profile of an olive ridley turtle headed directly toward the vegetation. A wide smile spreads across my face. We walk a bit closer and then stop at least 20 meters (65 feet) from the female. From here we have a good view and can monitor the stages of egg laying without disturbing her as she searches for the best spot to nest.

After a while, our turtle appears to have found a suitable place. Using her front and hind flippers, she begins to excavate a depression large enough to accommodate her whole body. Rotating on the spot while flinging sand out from under her, she creates what is known as a body pit. The Spanish description for this behavior is that she's "making her bed," *está haciendo la cama*. When she's satisfied with the result, she flattens and cleans a smaller area by wiping it with both of her hind flippers. Then she uses first one hind flipper and then the other to dig her nest and egg chamber. At this stage, she is less easily spooked, but to be on the safe side, my colleagues and I wait until she's halfway done with the egg chamber. Then we ready our research equipment and creep up on her from behind, still completely in the dark, to watch her lay her eggs. You might scare a nesting turtle if you approach her from the front, so this is something you should never do. And you should certainly never shine a flashlight directly into her face.

She pulls her hind flippers out of the nest and sets them down on either side of the hole. That is the sign we have been waiting for: she is now ready to lay her eggs. Some species cover their tail with one of their hind flippers right before they begin to lay. With what looks like considerable effort, our female squeezes her eggs out of the pair of oviducts that lead to her cloaca and from there they plop down into the nest. Right at the start we are counting up to three eggs dropping at once. It can take a turtle ten to twenty minutes to lay all her eggs. All the while, the mother-to-be is in a trancelike state. This is exhausting for her. She breathes deeply between pushes.

At this point I would like to mention that it is somewhat of a myth that the female falls into a real nesting trance as she lays her eggs, making her completely unaware of her

A size comparison between me and a leatherback mama busy camouflaging her nest highlights the majestic dimensions of the largest of all sea turtles.
ANDREY CASTILLO MACCARTHY

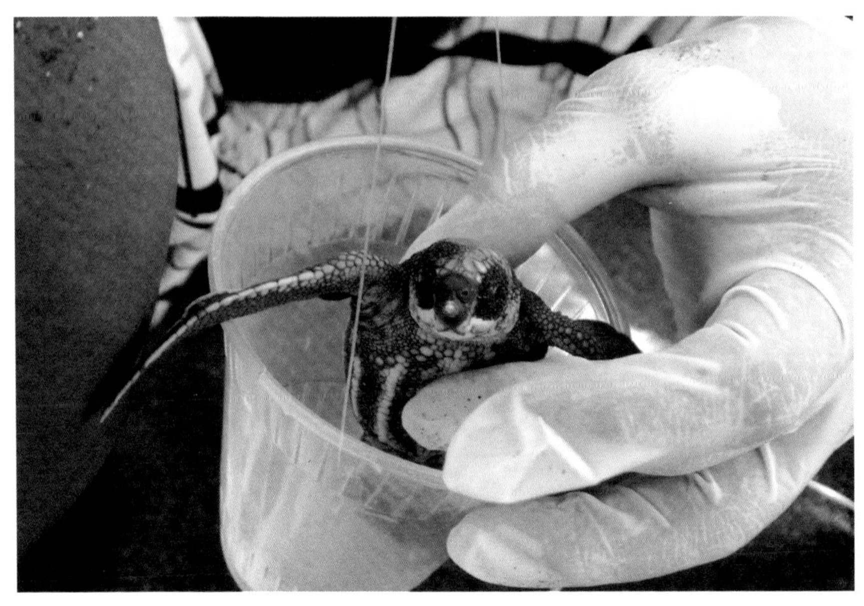

Weighing tiny hatchlings with what's available in the field. CHRISTINE FIGGENER

An out-of-this-world experience. During arribadas in Ostional, the beach is a teeming mass of thousands of olive ridley turtles that all lay their eggs at the same time. CODY GLASBRENNER

The transmitters we use today often weigh less than a bag of gummy bears, so light that the animals, like this olive ridley female, barely notice them. CHRISTINE FIGGENER

Leatherback eggs are round, about the size of a billiard ball, and protected by a soft, parchment-like shell. CHRISTINE FIGGENER

These leatherback babies are waiting to be released. CHRISTINE FIGGENER

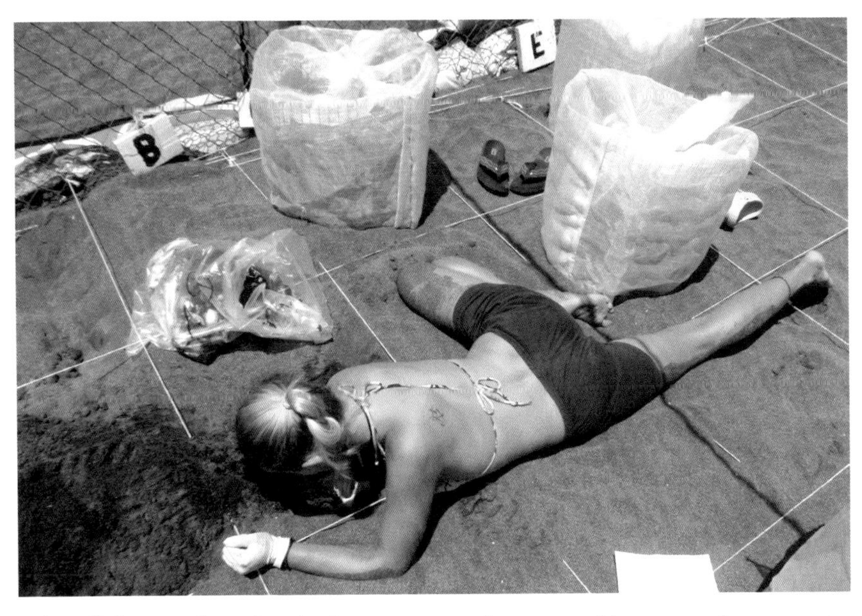

After all the eggs have hatched, we excavate the nest and inventory its contents. I carefully line up any unhatched eggs to open each one and find out what went wrong. LISETTE HEIJKE

I listen for my radio-tagged turtles with a directional antenna and a receiver to see if any of the females are nearby so I can catch them to take blood samples. BRIANNA MYER

When mating, the male uses his massive claws to hold on tight to the edge of the female's carapace—sometimes for hours. CHRISTINE FIGGENER

A green turtle enjoying its most important food source: seagrass. OLGA GA/UNSPLASH

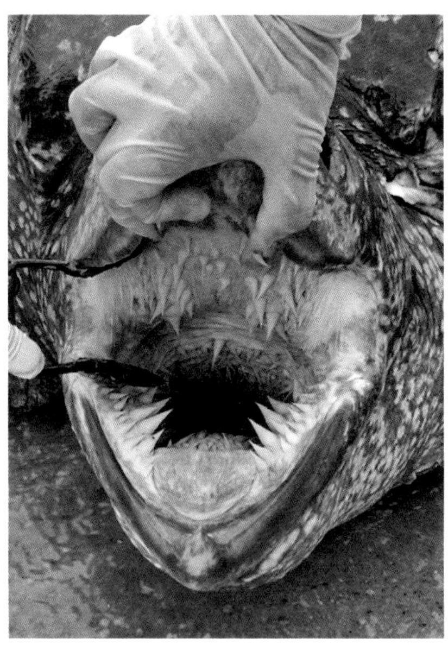

The inside of the mouth and the inside of the throat of a leatherback turtle are both covered in spines. TOM DOYLE

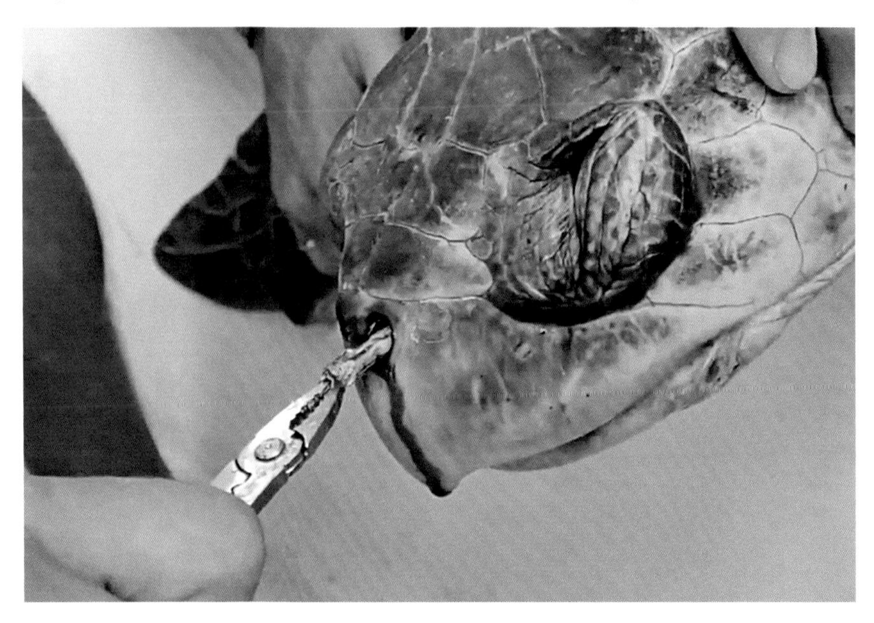

A screenshot from my viral video that documented how a member of my team removed a plastic straw from the nose of an olive ridley turtle.
CHRISTINE FIGGENER

A restaurant billboard from 1941 advertising turtle burgers. Fun for the whole family. MONROE COUNTY PUBLIC LIBRARY, FLORIDA KEYS HISTORY CENTER, ONLINE DIGITAL COLLECTION

A sea turtle fisher with a boat full of green turtles. Their meat was the main ingredient in turtle soup, a prized delicacy in the U.S. and Europe. MONROE COUNTY PUBLIC LIBRARY, FLORIDA KEYS HISTORY CENTER, ONLINE DIGITAL COLLECTION

People from around the world volunteer to help save sea turtles. This photo shows my research assistant Karen and volunteer Julia examining a newly hatched baby turtle. CHRISTINE FIGGENER

Introducing children to sea turtles early and educating them about the environment are the best ways to get the next generation excited about sea turtle conservation. ANDREY CASTILLO MACCARTHY

My friend and colleague Jairo, who was murdered on the Caribbean beach of Moín in 2013, with a baby leatherback in our hatchery in Ostional. CHRISTINE FIGGENER

My dog, Fiona. Since I adopted her from an animal shelter in Costa Rica, she has become my most faithful companion. CHRISTINE FIGGENER

surroundings. I, too, just used the word *trance* here and at the beginning of the book. But that's not exactly what happens. Most females do seem a little out of it and are less jumpy. For example, as I mentioned earlier, with leatherbacks and olive ridleys you can easily hold a bag under a female's cloaca to catch her eggs. You can also carefully take a blood sample and, in the case of a female leatherback, inject a microchip. But that is not true for all species of sea turtles or for all females, because even at this stage, a female could abort the nesting process. Hawksbills are known for being skittish, and it's best to leave them alone until they've finished laying all their eggs. Overly disturbed hawksbills have been known to continue to drop eggs while beating a hasty retreat back to the safety of the ocean.

Their "nesting trance," their awkwardness on land, and their lack of sharp teeth unfortunately make sea turtles and their eggs easy prey for poachers. Turtles can be snatched directly from beaches and there's not much they can do about it. The meat of the green turtle, in particular, is still highly prized in the Caribbean, where it's often prepared in a coconut sauce and can be found on menus, especially in smaller local restaurants.

Sea turtle eggs are also a popular ingredient in certain dishes in Costa Rican cuisine because they have a slightly different texture from chicken eggs. In many countries they are considered to be a natural aphrodisiac and are especially popular with older men. Of course, this is pure nonsense. Sea turtle eggs can no more boost libido or fix other dysfunctions than rhino horn or tiger bones can cure cancer. Despite this, you can find so-called sangrita shots offered at many bars: tomato juice, pepper, lemon juice, salt,

Panamanian chili pepper, a splash of orange juice, and, of course, a sea turtle egg. Luckily, it is illegal to sell or consume either sea turtle eggs or meat in Costa Rica and many other parts of the world. Despite this, the eggs are still a hot commodity on black markets worldwide and trade at US$1 apiece in Costa Rica. Depending on the species, one sea turtle lays between sixty and two hundred eggs per nest and nests from two to seven times a season. Egg thieves can rob many nests in a single night, making egg poaching a lucrative side hustle in a country where the official average wage is US$960 a month, but regular incomes in rural areas seldom exceed $15 a day.

The other highly prized commodity that sea turtles are still killed for today is tortoiseshell, which comes from the upper shell of the endangered hawksbill turtle. After the keratinized scutes of the carapace are washed and polished, their gorgeous pattern of amber flecks is revealed. This unusually flexible and malleable natural product was a predecessor to synthetic plastic. In Japan, for example, there is a centuries-old tradition of using *bekko*, the Japanese word for tortoiseshell, to make brooches, necklaces, combs, hairpins, eyeglass frames, guitar picks, and intricate filigree artwork. In the last one hundred years, millions of hawksbill turtles have been killed for their shells. In the twentieth century, Japan was the largest market in the world for tortoiseshell: government records show that from 1950 to 1992 more than 1.3 million adult and 575,000 juvenile turtles were imported.

The trade in tortoiseshell has a long history in Europe as well. More than two thousand years ago, Julius Caesar was amazed by the warehouses of Alexandria, which were filled to bursting with tortoiseshell—a welcome prize after his triumph over Egypt. Like the consumption of turtle soup, the trade in tortoiseshell really took off in colonial times. The European turtle fishery in the Caribbean ramped up from the middle of the seventeenth century in response to a steadily increasing demand for tortoiseshell. For three centuries, this once-abundant species was systematically plundered. When one of the local populations was wiped out, the turtle fishers simply moved on to the next one.

The trade in tortoiseshell over the past one hundred years is single-handedly responsible for the decimation and critically endangered status of the global population of hawksbill turtles. Today, researchers estimate that only a handful of nesting populations of more than one thousand females per year remain. The latest IUCN report for the species, in 2008, documents a sad decline in the global population of more than 80 percent in the last one hundred years. A study from 2019, based on data from the Atlantic, Pacific, and Indian Oceans, estimates that between 1844 and 1992 almost nine million hawksbill turtles were killed for their shells. As data on hawksbills from the Atlantic and Indian Oceans are far from complete, these numbers likely do not capture the full extent of the tragedy.

The trade in tortoiseshell has decreased since 1977, after the introduction of the CITES conservation agreement forbidding international trade in endangered species of wild fauna and flora. But illegal cross-border trade still exists. Prior to 1977, more than forty-five countries were involved in the trade of unprocessed tortoiseshell. When many important trading countries signed onto CITES, the number fell to just a handful of nations in the 1990s. The foremost among them, Japan, has still not outlawed trade in tortoiseshell and, years after CITES, was still receiving it from countries such as Panama, Cuba, and Indonesia. Finally, in 1992, Japan at least went as far as banning the import of new supplies of tortoiseshell. Since then it has, apparently, been satisfying internal demand from stocks it already has on hand.

And yet, in addition to the many other dangers they face, hawksbill turtles continue to be threatened by intensive illegal trade in tortoiseshell. Even though the same products can now be made from plastic, you can still find tortoiseshell jewelry sold at local markets and by street vendors in tropical lands. This jewelry is often sold under a different name. It is only when tourists reach customs at the airport that the awful realization dawns. To keep an eye on global trade and hopefully do a better job of combating it in the future, the organization SEE Turtles has developed an app called SEE Shell. People can use the app to identify and immediately report tortoiseshell products to highlight illegal activity.

Trade in sea turtle parts, like trade in ivory and rhino horn, is part of an active organized network that deals in wild animals across the globe. According to a 2012 report from WWF, this business is valued at an unbelievable US$20 billion a year, making it the fourth most lucrative illegal trade after the trafficking of drugs, humans, and weapons. Local poachers see little of this money, of course. Middlemen buy the goods and resell them at marked-up prices. Most poachers are in dire financial straits or live in regions where there are few to no opportunities for employment—but there are sea turtles. This means conservation efforts can succeed only when they consider the needs of the local population. At the end of the day, approaches that protect species and ecosystems but threaten the existence of the people who live in the region will fail. Everyone has a right to a life with dignity and an adequate livelihood—and that must be taken into account when conservation projects are designed.

Most poachers I know are nice people. They have children and play soccer on Sundays. They are nearly all men with no or only a rudimentary education. They have no steady employment but need to provide for their families. Over the years, the projects I worked for often hired poachers. Take Pablo, for instance. A wiry man in his late forties, he had a reputation for illegal hunting, stealing sea turtle eggs, and butchering green turtles. I often saw him on the beach, but the first time I spoke to him was at a funeral. He was polite and I was touched by the attention he was paying to his young daughter. One year I was short-staffed. After talking it over with the relevant government officials, I decided to offer Pablo a job. He would receive a salary for the eight months of the nesting season, social and medical insurance for himself and his family, and disability insurance. In exchange, he would patrol the beach for us, gather data, relocate nests—and, of course, he was not to poach any eggs or females or hunt illegally. Amazingly, my plan met with resistance in the village, of all places, where some people predicted things would not turn out well. They were wrong. Pablo became one of our best employees. He didn't drink, he was always on time, he was polite, and he possessed a wealth of knowledge about the sea turtles on the beach that made our job so much easier. I am convinced that no one had ever offered him the chance to prove himself before and he was grateful to accept ours.

It is in moments like these that I feel my work is really making a difference, not only for the sea turtles, but also for the people who live with them. It is not enough to patrol the beaches every night hoping to stop poachers from seizing eggs or the turtles themselves. The nightly patrols are only a Band-Aid solution for a much deeper wound. To tackle the problem in a systematic way, we have to remove the reasons for poaching, and the solutions can be different depending on the country. In Costa Rica, it means providing employment. Back in the early days of the 1980s and 1990s, many projects employed (former) poachers like Pablo, providing them with a regular salary. In most cases, this collaboration was successful and was a win-win-win situation: the turtles had one less poacher to fear, the former poacher had a secure income, and the project benefited from an additional worker who brought with him a wealth of expertise. I have learned an amazing amount about sea turtles from former poachers, things I would never find in textbooks and without which I would not be able to carry out my work with sea turtles nearly as well.

But this approach solves only one part of the problem. The other part concerns how much local people know about the threatened status of sea turtles and their understanding of the importance of sea turtles for the overall health of the ocean ecosystems on which their livelihoods depend. For example, in most coastal communities, the majority of animal protein that ends up on the table comes from fish, and people everywhere are noticing that stocks are dwindling. Ecosystems are complex networks of organisms that interact with each other, are intricately linked, and are mutually interdependent. Ecosystems are sensitive to change, and when whole species disappear, their ecological community is in danger of collapsing completely. Environmental education in local schools is, therefore, of the utmost importance. Children are extremely curious about the world around them. They are open to new knowledge, and they don't yet have to deal with adult problems. Educating children is the only way to make conservation sustainable. The children of today will grow up to be the adults who will call the shots tomorrow. I hope that by then they will have internalized the concept of sea turtle conservation so thoroughly that it will inform all their decision-making. These children are the future poachers—or protectors.

Although those of us who live in the Global North often see no need for any further incentives for nature conservation, this is far from the case for everyone else living on this planet. In many countries around the world, livelihoods are directly connected to nature. In contrast, people in Europe and North America mostly live cut off from the natural world and the forces of nature, and do not have to worry about how they are going to put food on the table tomorrow. As long as we have money, we can go to a store and buy everything we need to live and survive; we panic when threatened by a shortage of toilet paper. This makes it difficult for us to understand why nature conservation isn't embraced everywhere as having intrinsic worth, and that is also why conservation should offer financial incentives.

Our tendency to measure everything against Western values and our privileged positions is likely rooted in our colonial past. Even though those days are long behind us, neocolonial structures and ways of thinking are still a real problem in the world of conservation, which tends to be dominated by European and North American organizations and employees. As a white European woman, I cannot deny that I am likely part of the problem. We often feel we know best. We forget that we know what we know because we have seen up close and personal in our own countries what happens when you destroy nature. When we travel to tropical countries and see that people are in the process of making the same mistakes we did, we feel like the audience watching a play knowing what the next act will bring while the characters up on stage are none the wiser. We are stuck in patterns of the past and we adopt a condescending attitude as we try to teach other countries and people to do better than we did. Our own worldviews take precedence, and we ignore or downplay the needs of the local population. Even when we argue that we have "good intentions," this approach is still patronizing; it has been labeled the white savior complex. When you read between the lines, we are saying that local communities are incapable of making decisions in favor of conservation, and that they lack the ability to carry out necessary research and implement needed protective measures themselves. We see ourselves as white knights galloping in to save the day and forget that we are not omniscient, that some people do not want to be saved, and that, with appropriate support, they are perfectly capable of saving themselves.

It's sad to see how organizations, especially large organizations with money and influence, maintain this imbalance. They recruit staff from abroad to work in tropical countries, pay them European and North American wages, and exclude local professionals from promotions. As if that weren't unfair enough, they often pay locals much lower wages for the same work. Universities in the Global North are part of this problem as well. Scientists and students travel to tropical countries, help themselves to the local species diversity and even the assistance of local people for their research, but give nothing in return. We call this parachute science. Samples and data are taken home and nothing remains in the country. The local universities, government agencies, and communities, which could have benefited from the research results, are given a pat on the back as thanks and are often the butt of dismissive comments "between colleagues" back at home. Whenever I witness behavior like this, I am always on the verge of losing it. The casual assumption of superiority and the almost conspiratorial tone used to speak of the supposed incompetence of local scientists and assistants makes me hopping mad.

Most people in the Global North forget how privileged we are, even if we didn't grow up in rich families. Access to education and possession of a passport that allows us to travel to almost any country are at the very least two advantages many other people do not have. Most local scientists can only dream of the power and influence our nationalities give us, our research funds, and the state-of-the-art equipment we bring along. These advantages don't, of course, say anything about scientific competence. You can gather precise data with the simplest of tools. But we often forget that. Little wonder that local stakeholders are often extremely distrustful of North American and European scientists, and that bad experiences in the past often overshadow or even prohibit potential collaborations. That is a shame. Against this backdrop, I wholeheartedly support local people in their fight for independence, empowerment, and dignity. People want to be in charge of their own futures—everyone can understand that, right?

This is why I consider community-based conservation, developed in cooperation with the local community, as the approach most likely to succeed in the long term. One of the things that drives me is finding ways to empower the local population on the one hand and, on the other, to sensitize the next generation of conservation biologists from the Global North so they can explore alternatives beyond the white savior complex and neocolonial thinking. To realize these goals, I cofounded the Costa Rican Alliance for Sea Turtle Conservation and Science (COASTS) in 2014 with a few of my Costa Rican colleagues. In addition to undertaking research and protecting turtles, we are also working to change exactly these aforementioned things. Every year, students from around the world join our projects for a few months. They get to work with our local (imagine that) team of scientists and assistants, which allows them to develop a completely different level of appreciation for the knowledge and expertise of the people who live here.

KIM FROM THE U.S. is one of these students, and today she is accompanying Andrey and me on our patrol. She is sitting directly behind a female olive ridley, counting her eggs as they fall out. Andrey squats next to her and weighs ten of the eggs before he carefully lets them slip back down into the nest with the others. He explains to her exactly what he is doing and why, and describes to her how best to catch the eggs so they don't get covered in sand. This is Kim's first nesting turtle, and she's pretty excited. I find myself smiling. She reminds me of how I felt on my encounter with my first leatherback mama.

I'm pulled out of my reverie when I hear Andrey quietly whispering to Kim that the female has almost finished laying. The eggs are coming out of the cloaca more slowly now, which means that one of the two oviducts must be empty. As soon as all the eggs are in the nest, the female fills it with sand and then compacts it. All sea turtles do this using first one hind flipper and then the other, except for olive ridleys, which use their entire plastron to pack the sand on top of the nest. It looks like they are bopping to a beat only they can hear, and sea turtle biologists affectionately refer to the process as the "ridley dance." Watching them do this is great fun and always puts me in a good mood.

The end of egg laying is our signal to start collecting the rest of the data we need. The very first thing I do is check all four flippers for metal tags, the female's identity papers. I find some and note the numbers on our data sheet. If I hadn't found any identifying tags, I would have attached some using special pliers. These tags are vitally important because they are the only way we can get an accurate count of the females that come to our beach to nest. The international standard and most cost-effective way of identifying individuals is to attach two external metal tags to either the front or hind flippers. Microchips—passive integrated transponders or PITs—can be injected under the skin and, unlike metal tags, they don't fall off. Internal tags are somewhat more sophisticated, but these microchips are horrendously expensive. A metal tag costs about US$1; a

PIT costs around $8—and you can only read PIT tags if you shell out $500 to $2,000 for a scanner. In the Caribbean, we double-tag our leatherback females with two external metal tags and one PIT. In the Pacific, leatherbacks are marked with two PITs, which means only people who have scanners can identify them.

Next we record the female's size. Because her spine is part of her upper shell, we lay the measuring tape along the curve of her carapace to measure the complete anatomical structure of her back. This is called the curved carapace length. Baby turtles are so small that we measure only their straight carapace length, using calipers. An increase in growth rate indicates that the feeding grounds are productive and the female is getting enough to eat.

To complete our data collection, we look for unusual marks such as scars from boat propellers or fishing lines, or missing bits of flippers in the shape of a shark's mouth. That helps us get a better idea of the specific dangers our females encounter on their long journey to their nesting beach.

Now that the female has tamped down the sand above her nest, we watch as she starts camouflaging the area around the site so no marauders—at least no animal marauders—find her eggs. She spends about ten minutes throwing sand to distribute the smell of her eggs over a wider area, gradually moving farther and farther away from the egg chamber. This makes it more difficult for natural predators to track down the nest, but it does not fool human poachers. As females follow a set ritual when they nest, it's relatively easy for humans to find the eggs. They just need to know how to read the tracks. To hide the egg chamber from prying human eyes as well, we have to erase the tracks in and out of the body pit. An old poacher's trick!

If we need to move eggs, but we don't have a hatchery, we relocate them to a different, more appropriate location on the beach, where we dig an egg chamber of a similar size and shape to the original. Then we carefully place the eggs inside, one by one. We make sure they are still coated in mucus, which helps keep the eggs hydrated and is believed to keep fungi and bacteria from infecting them. When we're done, we carry any leftover sand down to the ocean and erase all our footprints and tracks so no poacher gets the idea this might be a good place to look for turtle eggs.

Someone waits with the female until she has safely made her way back to the water and disappeared under the waves. To return to the water, she orients herself by the lighter shade of the horizon above the ocean compared with the darker shade of the land, a navigation system that serves her well as long as the beach is undeveloped. Turtles both large and small are extremely sensitive to light and move toward the bright sheen of the sky reflected on the water. Developed beaches are a real danger to turtles. White streetlights and illumination outside—and even inside—buildings can disorient them, and they can end up crawling inland instead of toward the water. Coastal Florida, for example, is crowded with large buildings, hotels, and high-rises built right on the beach, all of which flood the area with light. The Sea Turtle Conservancy is one organization that has had great success over the years working with homeowners to install sea-turtle-friendly lighting with indirect light and red or amber light bulbs. Here in Costa Rica, we are fortunate that officially no houses can be built within the first 50 meters (165 feet) from the waterline, and development on the next 150 meters (490 feet) is strictly regulated. This means that most of our nesting beaches are pitch dark. Even so, a few females always seem to get lost. A few years ago, airport employees in Limón found a leatherback turtle on the runway. Someone had forgotten to shut off the landing lights overnight and they led the turtle astray. The fire brigade had to be called to transport the hefty animal—the turtle weighed at least half a ton—back to the ocean.

Another issue with developing and building on coastlines is the measures taken to protect shorelines from high water. For instance, the beach armoring that is supposed to protect houses from floods and storms robs sea turtles of nesting beaches. Tourism also complicates turtles' lives when sun worshippers occupy the beaches during the day, setting up beach furniture and ramming the tips of their sun umbrellas into the sand. Any eggs incubating under the sand are either squashed or impaled. Similar problems are caused by horseback riders, because horses are heavy and their hooves sink into the sand. Even the innocent pastime of building sandcastles and digging holes can have deadly consequences: when the baby turtles hatch at night, they can fall into the excavations and some are not able to make it out on their own. The next morning, they are easy prey for predatory birds or fry in the hot sun. So, when you take a vacation on a warm beach, it's always a good idea to check first to see if sea turtles nest there. If they do, we can all help make the beach safer for the nesting females, their incubating eggs, and the tiny turtle hatchlings. For example, you could destroy all sandcastles at the end of the day. You could also choose your hotel based on how sea turtle friendly it is. Does it use special lights during nesting season? Does it educate its guests about the presence of sea turtles? Does it bring in beach furniture at night and use weighted stands for its sun umbrellas? These small steps, which cost next to nothing, can really help sea turtles on their already arduous journeys.

It takes an olive ridley turtle about three-quarters of an hour to complete her entire nesting ritual; leatherbacks can spend up to two hours on land. You'd think that would be enough time to find the nesting turtles and gather the necessary data, but this works only if the sections of beach that each patrol has to cover are not too long. That's why we divide long beaches into sections that can be walked in less than an hour. But often patrols come across not just one female but several at once. That happened to me in the early days when I was patrolling a beach in Gandoca with my volunteer Sabine.

ON THAT NIGHT, having completed our first tour of our beach section, we are starting on our second when we spy a leatherback. We are right at the beginning of our section and the turtle is just coming out of the water, so we decide to walk on quickly and check the rest of our section before we begin to work on this turtle. We haven't gone more than 50 meters (165 feet) when a track up the beach alerts us to the presence of yet another leatherback, who is already "making her bed." I leave Sabine with a relocation bag—because the turtle is nesting too close to the water and we are going to have to move her eggs—and keep going.

I fall into a trot, and only 70 meters (230 feet) farther along I stumble upon the next track. This leatherback is also in the process of making her bed but luckily she is close to the beach vegetation and we can leave her eggs where they are. My adrenaline levels are rising. I run on and immediately see a fourth turtle, who is already digging her egg chamber. Holy moly! How many turtles do we have on the beach tonight? A few steps farther along the beach, I see a female who is already laying her eggs. I pull out my scanner to search her for a PIT. There is none, so I get out my injection gun and implant a microchip. Thank goodness she already has metal tags hanging from her flippers and her nest is in a good spot. I let the turtle get on with her business and take my two-way radio out of my pocket as I run to check the last half mile of our section. Gasping for breath, I contact my colleague Wilberth in the neighboring section.

"Will, Will, do you copy?"

"Go ahead!" I hear from the radio.

Out of breath, I sputter, "I've got five turtles here!"

"Whoa. What are they doing?" Will asks.

I had practiced the Spanish terms for the different stages of the nesting process with Will just a few days ago.

"One has just come out of the water, two are making their beds, one is digging her nest, and one is already laying eggs," I report excitedly in Spanish.

"Wow," I hear Will say.

He tells me he'll send his volunteer over right away. I run back to Sabine. Along the way, I check the fourth turtle, now also laying her eggs, for a PIT. She has a microchip but no metal tags. I can attach those later. I note the number of the PIT and hurry on. The third turtle has begun to excavate her egg chamber and I can see that she has metal tags. Noted. Check. As there's nothing more that can be done with this turtle yet, I hurry back to Sabine, who is now lying on her stomach catching eggs. I take my PIT scanner and scan this turtle's shoulder. No PIT. I inject one and run back to the first turtle, who has finally chosen a place to nest—unfortunately far too close to the water. She is now making her bed and so I have to wait before I can do anything. I rush a couple of hundred meters (about 600 feet) back to the fourth turtle, who has now finished laying. With lightning speed, I attach two metal tags to her flippers and measure her. Then back to the fifth turtle to measure her and back to the third turtle to check her PIT...

By now I am bathed in sweat and dashing madly over a stretch of about 500 meters (1,600 feet) from one turtle to the next so I don't lose any data or any eggs. Will's volunteer finally arrives and helps us gather the first turtle's eggs. This little game plays out for at least two hours until we have identified and measured all the turtles, injected PITs and attached metal tags as necessary, and made certain all the turtles are back in the water safely. I move the two nests where we had to collect eggs to a secret spot near the vegetation line and dispatch the two volunteers to make one more sweep of our section. I'm relieved when they don't find any more turtles. Completely exhausted and covered in sweat, we wait until midnight for the next shift to relieve us. When they innocently ask how things have been going, we break out into hysterical laughter as we hand over the backpacks with our research tools and data sheets. Pumped full of adrenaline, we walk to the village and review the events of the night over Snickers and beer.

AS A SEA TURTLE BIOLOGIST in Costa Rica, I walk an average of 12 to 20 kilometers (7–12 miles) a night, which saves me a visit to the gym. In the U.S. and Mexico, turtle projects can use quads on the beach, which of course makes it easier to cover more ground. That's not allowed in Costa Rica, and I think this is a good thing because motorized vehicles can severely damage beach ecosystems. Also, on the Caribbean coast, the beaches are so difficult to navigate that a quad wouldn't be much help anyway. Multiple streams that one would have to wade or swim across can suddenly appear, and the sand is covered in driftwood and other debris. We remove it regularly, but the sea spits out more so quickly we can't keep up. There's also little difference between high tide and low tide here, so I'm constantly walking through soft sand, scrambling over rocks and stones, and climbing up and down small dunes—a marathon that takes its toll as my pants become baggier and baggier over the course of a season. In comparison with patrols on the Caribbean side, patrols on the Pacific coast are a walk in the park. Tides iron the beaches smooth, the sand is hard, and you can walk for miles without much effort and without encountering any driftwood.

The biggest problem for sea turtles nesting on the Caribbean side is the erosion of some of the beaches. Since I started in Gandoca in 2007, we have lost more than 60 meters (200 feet) of beach. Fifteen years ago we had a wide inviting stretch of sand where we could build hatcheries to safely incubate our eggs for the duration of the nesting season from March to August. But in some places the sea now reaches as far as the jungle. It keeps nibbling at different spots and carrying away more and more of our sandy coastline. The most likely reason for this extreme erosion is that Costa Rica lies directly above the boundary between two continental plates that are constantly in motion. In addition, sea levels are rising because of climate change. Globally, sea levels in 2022 were on average 13 to 20 centimeters (5–8 inches) higher than back in 1900, and levels continue to rise every year. The Caribbean is a relatively shallow body of water, and a vertical increase of only a few millimeters (under a quarter of an inch) a year has a big effect on its horizontal spread. Naturally, this doesn't bode well for sea turtles, which depend on sandy beaches. Sea turtle biologists also appreciate wider beaches—if only because they are so much easier to walk on.

Wide beaches are especially important to ridley turtles with their distinctive arribada nesting behavior. They need lots of space for thousands of females and their clutches. Almost every month, the females all come ashore in the space of a few days to nest. Luckily most arribada beaches don't have problems with erosion. The problem they do have is that many clutches laid in the first few hours of this unique reproductive strategy are doomed when the next wave of females unwittingly digs up the nests of the earlier arrivals. And because olive ridley eggs take about forty-five days to hatch, the clutches must often survive not just one but two of these arribadas, which take place every twenty-eight days. In daylight, you can see millions of eggshells littering the sand of these beaches. Natural anarchy often prevails at beaches where protein-rich food sources are that abundant, Ostional in Costa Rica being no exception. Here, dogs systematically dig up eggs to binge feed on them, while groups of black vultures hunker down in the sand, waiting for their moment to pounce. What is not conveyed by the impressive photos and videos in the numerous articles and documentary films about arribadas is the distinctive, penetrating stink of rotting turtle eggs and vulture poop that hangs over the beach like an invisible veil. It is a smell I will forever recognize as that of Ostional.

Because of the enormous number of eggs that are laid here, you can find an average of three to four olive ridley clutches on 1 square meter (11 square feet) of beach, which is not particularly healthy for the individual eggs or for the nests. Consequently, only a small number of nests survive, and hatching success in individual nests—calculated as the percentage of eggs from which tiny turtles emerge—is extremely low. Depending on the season, it ranges from 0 to 32 percent and will likely drop in the future. The sheer number of rotting eggs turns the sandy beach into a gigantic compost pile. Decomposing eggs cause temperatures to rise and the developing eggs are starved of oxygen, which is used up by the microorganisms driving the decomposition process in the sand. That also affects the hatching success of the Eastern Pacific leatherback turtle nests that are laid here, which is a big fat zero. For many years now, we've been collecting each and every one of the leatherback eggs so we can move them to a hatchery with fresh, clean sand.

No one knows exactly why arribadas developed. This reproductive strategy must benefit female ridleys and their offspring or it would probably no longer exist, but much of this behavior remains a mystery. For example, we don't know whether this strategy is genetically driven, or if females can choose whether to participate in the synchronized mass nesting or to nest on their own. We also don't know what the life cycle of an arribada beach looks like. The overuse of the beach suggests a cyclical pattern leading up to a final collapse when hatching success declines due to the high density of nests. For example, according to accounts from villagers in Ostional, over one hundred years ago in the first decades the village existed, there were no arribadas. They came along later. And right now we are watching in real time as a new arribada beach is developing here in Costa Rica. For the past few years, smaller arribadas have been happening on the beach at Corozalito, and they are now increasing in size and frequency.

In the past Ostional suffered from massive problems related to the uncontrolled egg harvests that arose because of these arribadas and the millions of eggs that were laid every month. Villagers drove their pigs to the beach to fatten them up on sea turtle eggs, and hundreds of people from adjacent communities descended on the village to help themselves to the bounty. Fights broke out among villagers, and between villagers and the people who came in from outside. Because of this, in the 1980s, the community of Ostional decided to enlist government help to turn their beach into an official national wildlife refuge patrolled by rangers. They also asked the government to approve a legal egg-harvesting program that gave the village permission to dig up eggs laid in the first seventy-two hours of each arribada, distribute them to villagers for their own use, and sell the rest. They argued that these first eggs had a low chance of survival anyway and the money raised by selling them could be invested in protecting the beach by hiring guards who would make sure no one dug up the eggs illegally. The government agreed, and for decades this approach has been considered a textbook example of community-based conservation.

I can see the value of this program because at the time it was established, sea turtles were not yet protected in Costa Rica. But things have since changed, and there are aspects of the program that I find problematic. For example, there is no quota, no maximum number of eggs or nests that can be dug up. Limits are set only by the seventy-two-hour window and the number of people who are allowed to dig up the eggs. Since the program was put in place, the village has grown considerably, and the number of nests harvested has surely skyrocketed over the last few decades. Exact figures, however, are not made public. We also don't have good data showing whether the population of olive ridleys in Ostional has remained stable over the past forty years or whether it has increased or decreased. And so we have nothing to tell us how the program is affecting turtle numbers. In all the meetings and press releases, it's always reported that less than 1 percent of the eggs the turtles lay are harvested. The number of clutches harvested, however, ranges from a few hundred to a few thousand. According to a recent scientific study, only a few thousand females nest in the smaller arribadas from January to May, which can quickly raise the amount of eggs harvested to 50 percent.

Apart from that, I think that in the twenty-first century, we should not be exploiting wild animals over whose reproduction and fate we have no control, and the idea of exploiting a red-listed species for profit should not be passed on to future generations. A system of certification indicates that the eggs from Ostional, which are sold all over Costa Rica, have been harvested legally. But most Costa Ricans have no idea this certification exists or that it applies only to olive ridley eggs, which means the legal trade is opening the door for the black market. All too often the plastic bags with the Ostional seal are kept and filled with eggs from other species, and when I ask about the provenance of turtle eggs when they are offered for sale at market stands, I am told that they come "from Ostional" even though they are clearly not olive ridley eggs. For all these reasons, I am of the firm opinion that we need to end the consumption of sea turtle products of any kind.

Luckily, the zeitgeist is changing in Ostional and the community is developing more and more activities that don't depend on selling eggs to generate income. Tours are being offered that allow tourists to experience both the arribadas and the mass hatchings of babies that happen a few weeks later. The village also has a volunteer program where young people from around the world help collect data about the sea turtles and the arribada phenomenon. Another idea could be to incubate the supposedly doomed eggs in a hatchery instead of selling them, thus helping secure the future of the sea turtle population in Ostional. My hope is that these alternatives will soon replace the egg harvest, and that exploiting sea turtle species in consumptive ways will become a thing of the past.

Nesting is already hard work without all these additional obstacles and dangers. Simply laying eggs is exhausting, as you can clearly see from the turtles' increased rate of breathing, especially at the end. What with the energy-guzzling task of producing eggs, the arduous journey from the feeding grounds to the nesting areas, the grueling mating process with males, and laying up to seven clutches of eggs, females that start the season well nourished gradually become thinner and thinner. They eat next to nothing because the waters near their nesting beaches usually don't offer much in the way of nutritious food. The fat reserves they packed on the previous year need to last not only for the journey to the nesting beaches and back to their feeding grounds, but also for egg production and laying.

When the eggs finally lie in soft, warm sand and the mothers have disappeared once again into the dark water, the cycle is complete. Females and males have done their best to send another generation off on the same wonderful but dangerous life journey they themselves have undertaken and which has brought them back to this very beach. Now it's a matter of waiting and hoping that the turtles in this new generation will also find their way back—despite all the dangers.

– 8 –

THE LIFE OF A SEA TURTLE BIOLOGIST

IT'S OUR LAST NIGHT on the beach at Nancite, Costa Rica, and we're off on another night patrol. We are walking in single file along the edge of the waves when suddenly we spot tracks in front of us in the sand. I turn on my flashlight and recognize the paw prints of a big cat. "Jaguar," says my colleague Wilberth matter-of-factly. I look at him, my eyes round with amazement. "A jaguar has been on the beach with us all this time?" He nods and explains that the prints are, at most, a few minutes old.

I get goose bumps and my stomach drops to my feet. A few days ago, we found the remains of a dead sea turtle among the trees on the beach. It looked as though a jaguar had killed it. We've been feeling a feverish anticipation ever since, but up until now we have not seen one of these big cats face to face. My heart begins to beat faster and I trace a big circle over the beach with the light from my flashlight, hopeful but also fearful that I might spy a jaguar lurking in the darkness. With my flashlight beam lighting the way, we move on.

Suddenly a pair of eyes gleam from the darkness in front of us. Quite clearly cat's eyes. I stop as though struck by lightning and manage to croak, "Up in front, Wilberth." He directs his light toward the eyes as well, and they reflect the light in a greenish glow. It has to be the jaguar. It will probably clear off, I tell myself. Boy, am I wrong...

Framed as if by a spotlight, an elegant body emerges slowly from the darkness and makes its way directly toward us. Our lights have made the jaguar curious. I grab Wilberth by the arm and hold my breath. "Stay calm," he says quietly as he takes a couple of steps in the cat's direction. "Do you have your camera ready?" he asks. Without taking my eyes off the jaguar, which is now trotting toward us in a friendly manner as though it were a house cat, I take my camera out of my jacket pocket and turn it on.

Mere yards away from us, the jaguar comes to an abrupt halt. Its trusting curiosity has vanished and it is sniffing the cool night air suspiciously. It senses that something isn't quite right. I soak up this moment and stash it away in the deepest recesses of my heart. Even though I am almost peeing myself with excitement, I try to take note of every detail: its taut muscles visible under its gleaming coat; the wind off the ocean gently ruffling its silky fur; its long whiskers. What an amazingly beautiful, awe-inspiring animal.

I'm so enthralled, Wilberth startles me when he whispers "Now" into my ear. Reflexively I press the shutter release on my camera. The click scares the jaguar. In a single bound, it disappears into the bushes next to us. My legs are shaking and I take a desperately needed deep breath. Then I am overcome with a feeling of absolute euphoria. The encounter of a lifetime.

No matter how often I experience these miraculous encounters with nature, they never lose their special magic. I am eternally grateful for all the incredible and indescribable *National Geographic* moments I have lived through, and I understand why many children and teenagers dream of being (marine) biologists and working with charismatic species such as dolphins and sea turtles. But the reality of my work doesn't always match what most people imagine it must be like: lying under palm trees and playing with turtles.

My days look a little different now, but it wasn't so long ago that I had no fixed abode and moved every six months into wherever the headquarters happened to be for whatever project I was working on. I often shared a room with my assistants and sometimes also with a few short-term volunteers. It was fun at first. But when my husband, Andrey, and I began to work together on projects and we didn't have private quarters even as a married couple, it became decidedly less fun. Five days after our wedding we moved into a single room at the ranger station in Ostional with four assistants. Although today I barely remember what it was like, Andrey and I shared a tiny bunk bed. Every night we had a carefully choreographed routine so that we could both turn around at the same time without disturbing each other or stealing any more of the nonexistent space. We endured that for exactly one season. For the following seasons, I negotiated with my boss, who allowed us to rent a house in the village so that we only came to the station for meals and work. What luxury. Our own toilet and fridge—and no other people.

Life as a sea turtle biologist is far from glamorous. Usually our meals are provided by the project and prepared by a cook. It sounds like we are being spoiled, but what that really means is that we are served the typical Costa Rican dish of rice and beans three times a day. How it tastes depends on the cook's skill and how much money the project invests in the meals. I remember one year when corners were cut to save money and there just wasn't enough to eat; the mood at the station dropped to rock bottom. My complaints to the project's management fell on deaf ears until an ugly fight broke out over the last remaining pancake—and we were finally granted more food. Personally, I don't much like rice and beans, but even the biggest fan gets fed up after six months, which explains why good food was always a priority back then when I planned my day off—that and a fast internet connection.

Up until a few years ago, most projects had no internet connection and no cell phone reception. If I wanted to check emails, I had to get on the bus every couple of weeks and travel to the closest town, which was usually a good two or three hours away. Internet time was expensive, and I often had only a few hours until the bus returned. I spent most of my time reading emails, replying immediately, and loading up emails I had prepared beforehand. I focused on work communications. Exchanges with friends definitely took a back seat. Every once in a while, I sent a group email to everyone in Germany in which I described my work and my life—although my life at this time *was* my work and I didn't have much else to write about. One friend in particular took offense and officially ended our friendship. It hurt, but I understood her reaction.

It was very difficult for the people I communicated with to imagine my life in Costa Rica and to comprehend the difficulties I was facing at the time to keep in contact with them at all. My parents eventually resigned themselves to hearing from me only every few months. They embraced the maxim "no news is good news," and as long as they didn't hear anything from me, that probably meant all was well. Over the years, thankfully, opportunities to communicate improved and gradually more projects got hooked up to the internet, even if it was glacially slow. Cell phone reception became available in the villages, and I increasingly got at least one or maybe even two bars—as long as I positioned myself between the roots of the largest mango tree and held my arm up and out at the right angle.

It didn't help that we couldn't leave the project, because we didn't have a car in those days. We were completely dependent on public transportation, but that wasn't always available. And so we would spend more than six months in our tiny project bubble without any meaningful contact with the outside world until we changed locations for the next season, only to live in another bubble there. As the years passed, I had this overwhelming feeling that I was stuck and couldn't really engage with the big wide world.

I felt liberated, therefore, when I bought my car, Hercules, just a few years ago—what a luxury—and achieved a whole new level of mobility. My second luxury was that Andrey and I saved up and bought a small house with some land. It is in the same village where our organization, COASTS, operates a sea turtle conservation project from March to December. Thanks to this house and the coronavirus pandemic, after many years of upheaval, I have finally been able to settle down a bit and rest—even if our house also functions as storage for many items related to our projects and occasionally as accommodation for scientific assistants, and our land serves as a collection point for sorted beach trash.

Being so close to the project allows me to keep tabs on what is happening even if I no longer go out on patrol six nights a week. Because I work at my computer most days, I am pretty flexible and can schedule how much time I devote to hands-on work with the project, depending on how much I need to get done during the day. In theory, I get to patrol the beach with my team two to three nights a week, and more often at the beginning of the season when new people need to be trained. Reality of course often looks somewhat different because there is so much to do. Although I sometimes feel guilty about not going out on patrol as often as I used to, I enjoy being able to do my work well rested. And to be completely honest, I no longer have the physical or mental stamina to cope with a routine like the one I had for fourteen years.

These days, I wake up at five in the morning and go for a walk on the beach with my dog, Fiona, keeping an eye out for any turtle nests that have been missed or have hatched. Then I usually work out before sitting down at my computer to devote myself to my main tasks: administrative work for my consulting company, our conservation organization, and miscellaneous projects. I answer emails, take part in video calls, transcribe and analyze data, write scientific articles, file photographs, create posts for social media, organize events, pay bills, wire money, keep the books, search for new sources of financing, and a lot more. When I have time, I take part in afternoon beach cleanups and almost always supervise the excavations of hatched nests. The baby turtles remind me, especially at times when I otherwise barely see turtles, why I do the work that I do. I am happy that I also have some research and consulting projects that allow me to get out into the field for a few weeks every couple of months to get my regular sea turtle fix.

My transient life has made it quite difficult to make new friends over the past years. Most people I meet are just passing through, and as soon as I get to know them better and bond with them, they are on their way again. I usually do well being on my own and have gotten used to solving my problems alone, as my family and friends have been at the other end of the world for decades, but there have been times when I would have loved to just pick up the phone and call a close friend.

For that reason, my most important support system in Costa Rica has always been my sea turtle family, which is what I affectionately call my colleagues, students, and volunteers. As we are all separated from our own families, partners, and old friends for months, we become a surrogate family. Twenty-four hours a day, seven days a week, sometimes for six to eight months at a time, we share a bedroom and a bathroom, eat together, and organize whatever free time we have together. We laugh and cry with and about each other, share treats, and fight over dirty socks left on the floor. We get to know each other at our worst but also at our best, and through it all we are there for each other. We are bound together, whether we like it or not. Goodbyes at the end of each season are heartbreaking because we know that we often live on opposite ends of the earth and the chances that we will see each other again any time soon are slim. We stay in contact via social media, cheer each other on, celebrate our successes from afar, and often stay connected for life.

The highlight of the year for sea turtle biologists is the International Sea Turtle Society symposium, held at a different location each year. It's an opportunity to renew connections and swap stories. You could think of it as a mix of scientific meeting, summer camp, and vacation party. The conference allows us to catch up with the latest sea turtle science from various branches of research. We can talk shop as much as we want without anyone getting annoyed and rolling their eyes. We reconnect with friends and colleagues. Most importantly, we reconnect with our sea turtle family and share all the latest news.

LIVING AS I DO with an ever-changing cast of characters, I am aware of the advantages of having a partner at my side who is in the same line of work as I am. In the beginning, Andrey and I were often apart for weeks or months and sometimes reunited for no more than a day and a night before having to return to our respective projects. But after a few years, we were contracted together more and more frequently. Despite our differences, we make a good team as our knowledge base and personalities complement each other and bring more to the table than either of us could offer on our own. It is a great relief in our still-turbulent lives that we share a mutual understanding of our less-than-stable lifestyle and the demands it puts upon us every day—and also that Andrey accepts my preternatural need for independence. Any other partner would probably have thrown in the towel and said goodbye a long time ago, because my personal life is inextricably entwined with my sea turtles and I decide what I want to do and when I want to do it.

As a foreigner and a woman, I also find it helpful that the partner at my side is Costa Rican. Over the years, I have found it challenging to be a German woman in a country where patriarchal attitudes still prevail. On my very first day in Costa Rica, I noticed that I stuck out like a sore thumb. I felt like a giantess and my pale skin—even though I had a tan—hazel eyes, and blonde hair were all in stark contrast to the dark eyes, olive skin, and luxuriant black hair of my Latin American colleagues. I've always attracted a lot of attention from Costa Rican men, which made me feel extremely uncomfortable at first. After my experiences with German men, who are generally reserved even when they are passionate about something, I was not used to so much undisguised admiration or to being so obviously viewed as exotic. Men who are complete strangers sometimes ask to photograph me or secretly film me in the grocery store, just because I look different. I had a similar experience during my field trip in Egypt. Every time I went into a store, the owner asked if he could have his photograph taken with me. I also remember how the employees in the dive shop tried to get close to me before or after our dives. Innocent-seeming offers to help me into or out of my wet suit soon progressed to physical groping. That's why, to this day, my reaction is to pull back when men shower me with attention.

Over the years, I have gotten used to Costa Rican men's complete lack of self-censorship when they address women. I studiously ignore whistles from construction sites or catcalls from passing cars. The most they get from me is a tired smile. Although the men here are not usually physically aggressive, I don't want to gloss over the fact that sexual harassment does happen in sea turtle projects. This is unacceptable, obviously, but where people meet, problems are usually bound to happen. The most depressing thing in my experience is that such incidents have often been either ignored or downplayed by the mostly male project leaders. Even when they are taken seriously, the legal system in Costa Rica makes it difficult to report these incidents to the police, a symptom of the male-dominated world in which we still live.

As I've already mentioned, some of my colleagues also made life difficult for me at first. Comments like "You won't last more than two weeks here" gnawed at me, especially during the first few months. I began to doubt myself and wondered if I was really cut out for this work. Sixteen years later I can report without a shadow of a doubt that those voices were wrong. But earning the respect of some of the men was tough. It took endless discussions about responsibilities, childish pushbacks from the men, and a few crying sessions alone in the bathroom so I could vent my anger and frustration. Over the years, things became easier. I developed a thicker skin and learned how to avoid energy-draining arguments. I worked hard, kept learning, and got better at my job with every passing year until my critics had to admit that I wasn't as clueless as they once thought. It definitely helps that a new generation is stepping up today, one that has been raised differently and mostly doesn't have a problem being told what to do by a woman.

In the rural areas where many of my projects are located, I still attract attention and incomprehension, even after all these years. Many of the local people have spent their whole life in the same small village and just don't know what to make of me, a woman with not just a high school education but a university degree instead of children, who earns her own money, who is not financially dependent on a man, and who therefore calls her own shots. On top of that, I have a husband who cleans and cooks and does dirty laundry just as I do (we've both had to listen to ignorant comments on that score). People keep explaining to us that our decision not to have children goes against God's plan for man and wife. Sometimes it's hard, especially when the remarks are made by people I am close to. Naive comments from children are more amusing. When a friend of our then five-year-old niece Marylin asked me why I didn't have any children, Marylin didn't miss a beat as she explained to him that *white* women can't have children. That caught me by surprise, but she continued in all seriousness and with inarguable logic that she had never seen a white woman with a child. And indeed, every year hundreds of white women of all ages come to Marylin's village and not a single one has ever brought a child along. Marylin had thought things through and drawn her own conclusions based on her life experience.

The list of unpleasant and frankly dangerous situations women can get into through no fault of their own is long. A lot of these have happened to me in Costa Rica—after all, I have lived here for sixteen years—but such incidents are in no way restricted to Latin America or even to the rural areas where I spend much of my time. The scientific world is also full of misogyny, even though it might be less obvious. I am eternally grateful that over the course of my academic career, strong women have always stood by me offering help and support. By sharing their experiences, my female academic mentors helped me navigate the rocky path women in research must follow. They taught me how to deflect smarmy comments and avoid older colleagues known for "inadvertently" touching women on the bottom. Some of these situations could have had very real negative impacts on my career if I had reacted instinctively without thinking things through. In addition, there are the everyday annoyances of being a woman in science—being ignored or constantly interrupted during meetings, or being lectured by laypeople even when you are an expert in your field. We all become accustomed to these realities and often don't even register them as discrimination anymore. But precisely these things, both large and small, show that women are still not being recognized as equals.

Personally, I am particularly irritated by the mainstream bro culture among field biologists that completely excludes me as a woman. When I witness the casual ease with which some of my male colleagues present themselves, I sometimes wish that I felt the same sense of superiority, that I could have the same certainty I belong, that conviction the world is my oyster or at least owes me something, and that my spot at the table is a given. Of course, I can't know if this self-confidence is simulated or not, but I refuse to accept that it is based on superior male genes. I am firmly convinced that we, as a society, embrace the idea that male ego is more important than female ego. That is the only way I can explain the blindness of so many men. Do they really not notice that they monopolize conversations and seize every opportunity to promote themselves? Women are steamrollered and swept aside. How can we stand up for ourselves? Pushing back in kind seems easiest, and quite a few women react angrily and disrespectfully. I don't blame them, but I don't feel comfortable around so much negative energy. Feminism for me doesn't mean excluding others—including men. I don't hate men. I would just like more space, more recognition, and more respect for women.

When I talk about "women's issues" with male friends, they often roll their eyes. Many of them think the unequal treatment of men and women is a thing of the past. Yet women still put up with things men would never accept. This ever-present, subtle, and deeply rooted mundane sexism worries me. Women still have to learn how best to protect themselves in their daily lives, and what precautions they need to take in order to avoid getting themselves into difficult situations in the first place or, in the worst-case scenario, to avoid getting raped. I no longer have any inhibitions or shame about being proactive to prevent something bad from happening. But it's sad that women even need to navigate the world this way.

Unfortunately, many women, even extremely successful ones, have resigned themselves to the status quo or have no clue that they are textbook examples of internalized misogyny. "Luckily I'm not like other women" is a phrase even I used when I was young, and women who say this are part of the problem. They are part of the reason nothing changes. Instead of joining forces, we see ourselves as competitors for the few places set aside for us.

Fortunately, increasing numbers of men and women are becoming aware of the problem and are actively working for change. All of us—men, women, mothers, fathers—need to do our part to make this world safer for girls and women. I fervently hope that in the future women will come together in a more united front, do a better job supporting and protecting each other, and stand up for ourselves without fear even in the face of ignorance, scorn, or outright aggression.

For most people, my parents included, the idea that I, as a woman, go out more or less alone at night looking for sea turtles on isolated beaches likely makes their hair stand on end. To set these people's minds at ease: I almost always carry a machete and nowadays I'm usually accompanied by my dog, Fiona. Truth be told, I feel a lot less safe when I walk along streets in big cities at night. The risks that really worry me in my job and regularly lead to adventures—and, occasionally, emergencies—have more to do with adverse weather, remote locations, and impassable roads. There's usually no medical help on site and the nearest hospital is often many hours away, which can lead to some dangerous situations. I remember one in particular.

MY TEAM AND I are guests at a project on the Osa Peninsula, Costa Rica, where I am gathering data for my doctoral dissertation. I am in our *cabina* getting ready to go out on night patrol when my assistant Derek calls from outside: "Chris, we have a problem!" The urgency in his voice immediately puts me on edge and makes me suspect the worst. I grab our first aid kit before rushing out in his direction. Derek is sitting in the kitchen area with the project volunteer, Jasmin, and they are worriedly watching her foot slowly swell. Over the next few minutes, they tell me what happened. The young Swiss volunteer had been walking back to her *cabina* after the evening meal without a flashlight or sturdy shoes, when something suddenly bit her. She had the presence of mind to call out for Derek, who was still sitting in the kitchen reading. An experienced herpetologist, he immediately searched the area. Only a few yards from the path he spied a young fer-de-lance, a venomous snake in the viper family.

At the sight of the snakebite, I go on autopilot. In a heartbeat, I dig around in the first aid kit for a marker, which I use to draw around the edges of the swelling. I write down the time next to the outline. Then I get out a clean bandage and a sterile compress and wrap Jasmin's foot in a way that will allow us to loosen the bandage later if the swelling gets worse. Jasmin is amazingly composed. I instruct her to keep her foot below her heart and tell her the most important thing she can do is stay calm. After a short discussion, Derek and another assistant carry her to the project vehicle and the project leader drives her to the nearest town with an emergency room and a supply of antivenom. Jasmin is extremely lucky: the young snake has seemingly injected only a small amount of venom, and she doesn't suffer any lasting tissue or other damage.

Unpredictable weather and bad road conditions can also cause huge problems. I experience that firsthand one time when we are making our way back from the other side of the country. After a long, exhausting drive in torrential rain, a bridge and twenty more minutes on the road are all that separate us from a well-earned evening meal and a soft bed. I have crossed this bridge hundreds of times, but on this day, it has disappeared under about a foot of reddish-brown water. A man on a quad crossing the river from the other side assures me the water isn't particularly deep. If I had been unfamiliar with the lay of the land, I would have gotten out and waded through the river to test the footing. But in this case, I decide to trust my luck and drive over the invisible bridge. After all, I know where it is under all that muddy water, right?

I switch to 4×4 driving mode and steer Hercules to the spot where I believe the bridge is. I feel my tires making contact with concrete and am now convinced all will be well. I haven't gone more than a car's length when the right front tire slips off the bridge into the water. The car leans at a precarious angle with the left back tire now hanging in midair. I'm still not particularly worried. I shift into reverse to drive the car back onto the bridge. But nothing happens; my tires just spin. When the front tire slid off the bridge, the back right tire got wedged up against a massive concrete drainage pipe. I am now definitely up the creek without a paddle.

We are stuck fast and the footwell on the passenger side is slowly filling with water. "Quick, climb into the back and get out!" I shout to my front-seat passenger and colleague, Brie. She reacts with lightning speed, scrambling to the others in the back seat, climbing out with them, and wading to the bank. After thinking it over for a moment, I follow them. Our predicament brings out the Costa Rica I love so much. A number of men on motorcycles have been watching the whole spectacle, and they now rush to help. They are having an animated discussion about how to pull Hercules out of the river when a pickup truck drives up. The driver wants to turn around as soon as he sees the flooded bridge, but then he becomes aware of our situation. The four men in the vehicle immediately offer to tow Hercules out if we can rustle up a tow chain. One of the motorcyclists roars off to find one. Meanwhile, I try to reassure my students. "It's no big deal," I tell them. "Help is on the way." Although I am calm on the outside, I am having a hard time convincing myself as I imagine Hercules being washed down the river with all our equipment and personal belongings inside.

When the tow chain finally arrives, I crawl back into the driver's seat and Hercules is pulled out of the river without any problem. Unfortunately, the electronics in the footwell on the passenger side got wet, and my trusty car won't start. The men in the pickup offer to drive me to a mechanic in the nearest village. As I climb into the passenger seat and we drive off, I see the anxious look on Brie's face. I can read the thoughts going through her mind: "Hopefully you won't get raped, cut up into little pieces, and buried somewhere in the jungle." I have to admit I am sincerely hoping that Brie or one of the others has at least made a note of the truck's license number. But the four men—a father and son and two employees—are extremely respectful and super nice. Not only do they drive me to the mechanic, they also buy beer for our whole team even though I feel I should be the one buying them beer to say thank you. But the father waves off my offer and says, "You guys need it more than we do!" Then they drive me back to the river to wait for the mechanic. When I return, I think I can see a few tears of relief in Brie's eyes.

At this point, I'll say, "Hi, Mom!" For many years I have lived by the maxim "What Mom doesn't know doesn't worry her," and I usually don't tell my parents about experiences like this one so they won't worry unnecessarily. However, as my mom will likely read this book and as she has probably already taken a few deep breaths, I'm going to say, "Sorry, Mom, but at least I am still alive."

In my line of work, negative and positive experiences, moments of wonder and times of deep despair, the fate of the world and personal tragedy are never far away. We are working for the greater good, but we are still human beings with our personal worries, needs, and emotions. So it is not surprising that the toxic situations I am often exposed to through my work, which I could avoid under different circumstances, are chipping away at my mental health. Over and over again I have to stare into the abyss of human existence and try not to let it swallow me up. I would love to erase from my memory many horrifying things I have seen people do—to sea turtles, to other animals, to their fellow human beings—because when I think about them too much, I lose faith in humanity and sincerely question our right to exist as a species.

One of the saddest and most traumatic moments in my life was when my friend and colleague Jairo "Foca" Mora Sandoval was murdered on a beach on the Caribbean coast of Costa Rica near Moín in 2013, when he was just twenty-six years old. He was killed by alleged poachers. Jairo grew up in Gandoca, my husband's hometown and the place where my passion for sea turtles began. Jairo had been part of my sea turtle family for many years. We first met in Gandoca through Andrey and the work our sea turtle project was doing there. In 2010, Jairo accompanied us to Ostional for the first time, and we researched and protected sea turtles together there for many seasons. Jairo grew up in modest circumstances with three sisters, and his father raised cattle. What made Jairo stick out was a combination of intelligence, inquisitiveness, and passion for conservation. Over the years, he blossomed into a competent leader who brought with him the knowledge of a local who had grown up with sea turtles on the beach—a rare mix we always keep an eye out for in people but seldom find. Even though Jairo never went to college and got his high school diploma just a few months before his death, I had great hopes that one day he would be one of the leading figures in sea turtle conservation in Costa Rica. Unfortunately, things turned out differently.

The day Jairo was murdered is burned into my soul. I was visiting Germany and sitting in my grandma's living room looking out into the garden, where a pair of great tits were feeding their newly hatched chicks, when the telephone rang. My boss from Costa Rica was on the line. "The news I have is not good. Jairo didn't come home from the beach last night." Only the four young women who had been out on patrol with him had returned to the station, completely distraught, without Jairo or their car. The alarm was raised immediately, the area was searched, and a body was found. At this point my boss didn't know if it was Jairo, but he feared it was. Andrey's brother was on his way to identify the corpse.

I listened, frozen with shock. I simply could not grasp what had happened and I was certain it must be a mistake. But a few hours later our worst fears were confirmed. It was as though the ground had been swept out from under my feet. I was having difficulty accepting the finality of this dreadful news, while at the same time trying to comfort Andrey. Hadn't I just chatted with Jairo a couple of nights ago as I was helping him book a flight to Germany? It simply could not be true. Despite all the dire situations and adventures, in the early days especially, I rarely thought about our work being seriously dangerous and potentially fatal. When you're young, you think you are immortal. And you can convince yourself of that until something happens to remind you that life can be over in a flash.

Jairo's murder attracted international attention, and his name joined a long list of murdered conservationists. A few large conservation organizations and individuals used Jairo's death to promote their own agendas, including pointing fingers at Costa Rica, or to drive up their viewer ratings. For those of us who were grieving the loss of a good friend and family member, this was hard to watch.

But we were not the only ones to suffer under the negative media hype; it also hurt tourism in Costa Rica and hit sea turtle conservation efforts particularly hard. The beach at Moín is a special case. It is right by Costa Rica's largest shipping port and is not part of one of the country's many protected areas. Large port cities attract a lot of criminal activity and organized crime, and surrounding areas are often precarious places. Reports of Jairo's death, however, made it sound as if suddenly all the beaches in Costa Rica were extremely dangerous and the villages were human cesspits. This, of course, was completely false and the reporting did many people and regions a great disservice. Tourists were suddenly scared to travel to Costa Rica, and they certainly didn't want to work for sea turtle conservation projects. But most of these projects are funded by what is known as voluntourism (a form of ecotourism combining volunteering with tourism), meaning that members of the public pay to work for conservation projects in return for spending their vacation time in a more meaningful way. This is how the projects can afford to cover their expenses and pay local assistants. When the tourists stayed away, many of these projects found themselves struggling to survive.

Then there were those people who suddenly embraced Jairo as their personal persecuted savior and wanted to confront the "evil" poachers with violence and AK-47s. Their ideas drove deep rifts between the villages lying close to sea turtle beaches and the conservation projects working there. Some of these projects had spent decades gaining the trust of local residents and had worked hard to find conservation measures and solutions that benefited both sides. Relationships built on trust and people's life's work went up in smoke overnight. The rabid cowboys who wanted to protect sea turtles by force and thought that would make Jairo's mission live on disappeared as quickly as they had come, leaving many organizations with nothing but the shattered remains of their projects.

Four of the seven men who were accused of Jairo's murder were finally convicted in 2016 after an initial mistrial. They are currently in prison. I, however, find neither relief nor joy in this because their convictions will not bring Jairo back. And many questions surrounding his death remain unanswered. Some people believe the men were involved in the drug trade; others are convinced that they were hired by the company developing the big new shipping port in Moín. In both scenarios, the motivation was to drive away the sea turtle project to have free rein for their own plans. We will likely never know why Jairo was killed.

AFTER JAIRO'S DEATH I found myself in a dark place and struggled with questions about my purpose in life and about the universe in general. I had never felt so powerless or lost. After a few months of aimless drifting, I looked deep within myself and wondered what I could do better and how I could help my sea turtles more effectively in the future as they travel through the vastness of the ocean from their feeding areas to their nesting and mating areas and back. For a few years by then, living as I did tucked away deep in the Costa Rican jungle, patrolling remote nesting beaches at night, I had been feeling increasingly cut off from the outside world. I felt that life was passing me by. Surely I could do more. Indeed, I had to do more. I owed it to both my sea turtles and to Jairo. And so I decided to return to university for a PhD to expand both my network and my expertise. I wanted to have more independence and a greater impact on the global future of sea turtles.

The universe granted me my wish but in a different and totally unexpected way, when in 2015 I filmed the removal of a plastic straw from the nose of an olive ridley turtle. In hindsight, I could not have asked for a better way to expand my circle of influence, reach more people, and set something in motion—even though this all happened because of an extremely disturbing discovery. Thanks to this one video, I acquired an eclectic mix of skills that had nothing to do with science, ranging from expertise in copyright issues and social media marketing to unconventional methods of fundraising. I probably also got a thicker skin from having to deal with blunt anonymous and not-so-anonymous criticism, and rude and ignorant remarks. Nameless internet users attacked me, as did powerful lobbyists who made outrageous claims as they attempted to drag my name and my video through the mud.

The most important takeaway for me was that those of us who are scientists need to communicate our findings in a more accessible way. Rather than simply publishing them in scientific journals, we need to use social media and the press. Unfortunately, many of us are not good at doing this or are not interested; for others, this is uncharted territory. Yet others have a deep mistrust of and unease about popular media because of the many simplifications and resulting inaccuracies in the stories they tell.

In recent years, despite misgivings, I have ventured further and further into science communication even though I have no background in this field and I loathe talking to unfamiliar people. It stresses me out, and every interaction takes a huge amount of effort. It feels strange to suddenly stand in the spotlight having to worry about how I look in front of a camera. Field biologists spend long days doing dirty, smelly work, often with no time to even brush their hair. We are anything but glamorous, and I often feel out of place next to television presenters and professional influencers. I believe many scientists shy away from the media for exactly the same reason.

The ground rules are also very different. Media run on emotions; in science emotions are taboo, which might be why scientists often come across as cold or boring. But if we succumb to the emotional language of the media, many of us are afraid that we will lose credibility with our colleagues and our funders—a not-unfounded concern. Although we are well aware of what is going on with nature and animals, as scientists we are expected to take an objective stand and not express any personal opinions. But we are not automatons, we are people—people who are concerned about what is happening.

For this reason, science and activism should not be seen as mutually exclusive, as they are often portrayed to be. Years ago, two of the most famous sea turtle biologists, Dr. Nicholas Mrosovsky and Dr. Peter Pritchard, warned that when we are looking for solutions to conservation issues, the role of science should not be undermined by the passion of activism, no matter how seductive it might be. I agree that, then as now, feelings have no place in the scientific process. Facts, not emotions, should lead conservation efforts. However, I am also of the opinion that scientists can no longer stand silently behind their data and watch as the human race destroys everything on which our lives and those of many other organisms depend—precisely because scientists know and understand a great deal that most people might not even see.

Activism and science are truly effective only if they work hand in hand. We need accurate data to inform and direct activism. Dramatic videos are good to generate attention and actions, but the message behind them must stand up to scrutiny. If it doesn't, the label of being frivolous and sensational will stick, as well it should. I realize that constant vigilance and care are necessary when scientists interact with the media to avoid misinterpretation and misinformation. Over the years I have learned that reporting that touches hearts is probably the most effective tool to convey critical information, not just to fellow citizens but also to political decision-makers. By shaking them up, we advocate for the conservation and survival both of the environment and, by extension, of humanity. Despite this, I am first and foremost a scientist. My data are what count, no matter how I feel about them. I can, however, decide how to tell their story.

After my viral video, I consciously began to use the platform it created for my work and my messaging about sea turtles because social media has great power and a wide reach when it comes to forming opinions. Finding novel ways to share interesting and convincing messages provides me with both a challenge and a creative outlet. Telling the public about my sea turtles and the threats they face is important to me. I hope I am raising sufficient awareness that it will lead to concrete conservation actions. I answer emails from children who want to reduce the use of plastic in their schools or neighborhoods. I now get invited as a guest on podcasts and can talk as much as I want about what is closest to my heart: sea turtles and the threats they face. I have taken part in a few documentary films and campaigns fighting for the ban of various plastic products. I have been invited to give presentations at schools, conferences, and other events. A highlight was when, to my surprise, *Time* magazine named me as a Next Generation Leader alongside celebrities such as Ariana Grande and BTS in 2018. When I got the call, my first reaction was to ask them if they had made a mistake and meant to bestow this honor upon someone else. What a mad world we live in. I was still as proud as Punch!

Fortunately, times are changing, and Costa Rica is no exception. Thanks to the coronavirus pandemic, internet connections and cell phone reception have even made it as far as the small village where I live in the southern Caribbean. Whereas sixteen years ago I had to make calls from the only public telephone box in the village and spend three hours on the bus traveling to distant Puerto Viejo every two weeks, today I can sit in the comfort of my own home and communicate with people all over the world. This alone helps me feel I am no longer cut off from the rest of the world.

Many people who have only a vague idea of what I do and know nothing of the hardship and struggle, who see only sun, ocean, palm trees, and adorable baby sea turtles, ask me how I managed to pull it off. Some probably want to convince themselves that I'm just lucky to live what looks like a charmed life. I vehemently disagree, because it has been anything but easy and I worked hard for what I have and what I am able to do. The truth is that sometimes even I don't know exactly how I managed to realize my dreams. I can't give anyone a template to follow. The way I set goals and reach them changes from day to day, week to week, and year to year.

I can certainly chalk some of this up to my resilience. I'm a bit like my sea turtles in this respect: when I get knocked down, I get right back up again. If something goes wrong and everything seems hopeless, I sleep on it, and by the next day the world looks completely different. I hardly ever give up. My ability to be flexible and adapt, my desire to learn, and my trust in my own abilities probably also contribute to my success. Over the years, I have had to keep adapting to different locations and different people, learn new things, overcome obstacles, fight bitter battles with myself and with others, and sometimes even admit defeat. But every time, every single time, I had and still have the privilege of learning something new so that I could continue to grow.

Thanks to my work I have seen and experienced the most amazing things. Sunrises and sunsets in dazzling colors. Encounters with animals that other people know only from television. The wildest ocean adventures and the most beautiful journeys. Of course, there were and are all those unpleasant moments: witnessing atrocities, experiencing frustration, having no fixed address, and moving every six months. But I concentrate on the good things in my life. When I hit a brick wall, I pave my own way and tune out the ridiculous chatter of pessimists. That has allowed me to shape my life the way I want. I know that in the eyes of some people I am not particularly successful: I studied the wrong subjects, earned my doctorate too late, and didn't pursue a career in academia. I don't own a house in a fancy new neighborhood with an expensive sports car in the driveway. But perhaps we should start defining success differently. For my part, I will die content if I know I have achieved something positive with my life. As my grandma always told me, "You can't take money and riches with you anyway."

By this point one thing should be clear: my work is not a nine-to-five job. If you want to go into nature conservation to play with sea turtles and dolphins, enjoy warm Caribbean nights, and be well paid, I strongly suggest you rethink your career choices. The sacrifices you will have to make and the toxic situations you will find yourself in time and again call for courage that can be sustained in the long term only if you are passionate about your work and if you believe it will make a difference.

For those of you who feel exactly this kind of passion, who are convinced that they are fighting for a higher cause but are being held back by those around them: Ignore the negative comments! They mostly come from those who never dared to live their dreams. I want young people to see that there are different ways to realize visions and achieve goals. Nowadays the world is our oyster, and even in our own countries there is more than one place to be happy.

We just have to find what makes us happy, what we are passionate about, what our purpose is. Easier said than done, but maybe we should spend more time finding meaning in life than making money. What do I mean by that? Don't let naysayers get you down. Go on your own journey!

– 9 –

A NEW GENERATION

WHEN I WATCH baby sea turtles hatch and see their little beaks with their egg tooth emerging from the nest, then watch the hundred tiny pairs of flippers moving in unison as the babies break through the sand and make a mad dash to the sea, I cannot help but be happy and rejoice in a feeling of accomplishment. Despite all the dangers, a new generation has made it to the water to begin their life journey.

For at least two decades, in the 1970s and 1980s, almost 100 percent of sea turtle nests on the Caribbean beaches in Costa Rica were plundered. This meant, of course, that a vanishingly small number of babies hatched, which resulted in a generation gap in today's population. Given the longevity of sea turtles, that gap did not become obvious until the 1990s. Thanks to laws being passed and various new conservation projects being established, the plunder declined significantly, and new generations of sea turtles that have since hatched now have a fighting chance.

I am happy to report more good news and developments in sea turtle protection in other places over the past few years and decades. For example, since the 1960s, the population of Kemp's ridley sea turtles was in steady decline until it reached a record-breaking nadir in 1985 of only 702 nests in the Gulf of Mexico, their main habitat. The decline was due to the destruction and poaching of 90 percent of the turtle nests the very same day the eggs were laid. There were also the previously mentioned enormous slaughter-houses in Mexico, where ridley turtles were killed so their skin could be used to make leather for boots. The shrimp fishery with its trawl nets also played its part. And so, in 1978, government representatives from Mexico and the U.S. got together in a last desperate and ambitious attempt to protect the species.

The idea behind the Kemp's Ridley Headstart Project mentioned earlier was to protect nesting females, their eggs, and juvenile turtles on their main arribada nesting beach in Rancho Nuevo, Mexico, while also establishing a secondary nesting colony in Texas. Some of the nests laid every year in Rancho Nuevo were collected and moved to Padre Island National Seashore, a nature preserve in Texas. As soon as the clutches of eggs arrived, they were incubated in climate chambers under controlled conditions. The freshly hatched turtles were released on the beach to crawl down to the waves on their own. After a short excursion into the Gulf of Mexico, they were then recaptured with nets and raised in tanks for a few years to increase their size and, with that, their chances of survival. The idea was to imprint the young animals on the National Seashore so they would return there as adults to mate and nest.

Meanwhile, laws were enacted that required all shrimp trawlers along the Texas coast and in many other areas to install turtle excluding devices (TEDS) on their nets to prevent the incidental bycatch of turtles. In addition, beginning in 2000, the Texas Parks and Wildlife Department started to close the coastal waters of southern Texas to shrimp fishing from December 1 to May 15. The years-long coordinated efforts paid off and led to a substantial increase in the number of ridley turtle nests found in Texas and Mexico from 1997 to 2009, with a high count of almost two thousand nests in 2009. That is only one of many examples of how valuable cooperative efforts with a common goal can be and what we can achieve when we collaborate.

In most of my projects, between twenty thousand and fifty thousand baby turtles hatch every season. They form the basis for a new generation even if many of them never reach sexual maturity. Thanks to our tagging efforts, we know that 10 to 20 percent of our nesting populations are first-time nesting females, so-called neophytes, and this number has been increasing. However, that is not enough to increase the size of the population. For this to happen the numbers must be higher than the mortality rate for sexually mature turtles. The overall population trend is, unfortunately, sloping downward, even though at a lesser rate than maybe twenty years ago. This is likely due to the intensive conservation efforts on many nesting beaches combined with new regulations for fishing gear. I still have hopes that sea turtles will exist ten, twenty, or fifty years from now, if we keep helping new generations to hatch.

And to do that, we also continuously need new generations of sea turtle conservationists. Despite all the hardships, our work also offers many rewards. In addition to feeling good about helping sea turtles, we experience breathtaking moments and are part of a dedicated global community. Luckily, an astonishingly large number of people love sea turtles as much as I do. Perhaps it is because, like dolphins and pandas, they are adorable and not at all threatening. In professional circles, animals with these characteristics are called "charismatic megafauna." This is in contrast to animals we find harder to love but that are also in need of our protection, such as snakes, sharks, or the lesser known blobfish. Sea turtles, therefore, got lucky. Suspended in a world just beyond our reach, they are surrounded by an aura of mystery that sparks our curiosity.

This fascination motivates many vacationers, especially young people, to want to be a voluntourist in a sea turtle conservation project rather than another sunbather on a beach. Our volunteers come from countries all over the world and all sections of society and—perhaps surprisingly—not all of them are studying biology. In fact, the minority have a full-time career in conservation. What all our volunteers have in common is a passion for the ocean and sea turtles and the desire to contribute to their protection and survival. In our projects, they are rewarded with the sight of sea turtles in the wild and they can personally accompany them on parts of their life journey, beginning with the hatching of the babies and ending with the nesting of the females. For many people, this is a one-of-a-kind experience.

These volunteers have had an enormous impact on my work, and in years past they were often my only contact with the outside world. Before the pandemic, many hundreds of volunteers often participated in my projects for a week or two or even for a few months. Over the years I have had the privilege of getting to know many amazing, caring people, all of whom are fighting for a better world—specifically for a future for sea turtles on this planet but in the end for all of us. Many of them are just starting out on their life journey. They are high school graduates who are either taking time off before going to college or have already embarked on their studies. A smaller number are people who have arrived at a crossroads and want to ground themselves and rethink their lives. And finally, there are the retirees traveling the world and wanting to make a difference, like Paul from France, who at the age of eighty-five spent many weeks patrolling the beaches with us.

All these people inspired and continue to inspire me, and many left lasting impressions. I became an adult in their company and some of them became lifelong friends. I recall hundreds of amusing anecdotes and many experiences I connect with individuals I cherish.

There was my assistant Cody, who once had to answer a call of nature and disappeared into the darkness under the trees. Suddenly we heard him shriek and curse as he ran back in our direction. As he approached, we were hit by the unmistakable smell of skunk. Cody had inadvertently peed directly on one of these sweet little fellows. It took offense at being so rudely awakened and immediately sprayed Cody. Luckily, the skunk only caught Cody's pants, which he could remove; otherwise he would have had to spend the night outside.

Then there was my assistant Cybil, an experienced long-distance runner, who had already participated in a number of marathons. She could run effortlessly on sand. One night we heard over the radio that a leatherback turtle had crawled out of the water at least 4 kilometers (2.5 miles) away at the other end of the beach. Cybil jogged behind me with our research equipment on her back without a whisper of complaint. When we arrived at the turtle, we discovered that the zipper on the backpack had opened while we were running and pieces of equipment had fallen out one after the other all along the beach. Cybil found this quite embarrassing and she ran back the whole distance along the beach to gather up our things in record time.

And then there was Jason, a volunteer who was helping me with a leatherback turtle. He was holding one end of a thermocouple cable while the other end was tucked in among the eggs to measure the temperature of the nest. I instructed him to keep a firm hold of his end of the cable while the female filled in her nest with sand. On no account was he to let go. A large group of tourists was in attendance that night and I backed off so as not to obstruct their view. I was chatting with one of the guides when another guide suddenly tapped me on the shoulder. "You know your volunteer is still behind the turtle, right?" Damn. I had completely forgotten. As I rushed toward him, I couldn't help but laugh out loud. Jason was now covered with sand from head to toe and was even sporting a little peaked cap of sand on his head. He had kept hold of the cable as instructed, but what he should have done was tie it to a branch as soon as the female finished filling her nest and began camouflaging it by throwing sand all over the place. Huge amounts of sand get thrown everywhere when a turtle camouflages her nest, which can quickly bury any rain poncho or flashlight—or volunteer—carelessly left behind her.

Without these volunteers, many projects would not be able to cope with their workload. You can only monitor hatcheries 24/7 and send patrols out onto the beach all night long if you have enough people. Volunteers also help us construct hatcheries at the beginning of the season, prepare the beach for gathering data, and pitch in with the never-ending task of picking up plastic debris. Many things would not be possible without our volunteers. A good team of volunteers really can move mountains. These additional team members can also be a breath of fresh air.

Volunteers are also my constant companions on the nightly patrols, most of which last four to five hours. Much of the time is spent in conversation to keep us awake, especially when no sea turtles are nesting. Darkness and exhaustion make it easier to talk about difficult and emotional subjects and to philosophize about God and the world. There is something otherworldly, almost magical, when you are hypnotized by the sounds of the night, walking along the beach under the infinite canopy of stars, or taking a break and just sitting there listening to the waves. You can allow your thoughts to wander and the world seems clearer. Many of us often think about the future on nights like these. The future of our beloved sea turtles, but also our own futures and the future of all humankind. Plans are forged and ideas developed, but above all, people find themselves. It is perhaps not surprising that the weeks and months they spend in our projects are life-changing for many volunteers. In 2008, for example, when I was studying for my master's degree, I got to know Friedi Nolting , Sonja Renschler, and Tino Scraback in Gandoca. The three of them were in Costa Rica for three months to learn Spanish, explore the country, and get to know its people while doing something meaningful. To this day, they are all still active in various environmental groups, and in 2020 they founded ProMar e.V., a nonprofit organization in Germany that supports my work in Costa Rica.

Naturally, peace, joy, and harmony do not always prevail. The meager financial resources most projects have access to, and the resulting lack of paid professionals, mean positions are often covered by volunteers, which leads to a certain inconsistency in the quality of the help. When you depend on volunteers, you never quite know what you are getting. You can get absolute gems but also obnoxious companions lazy enough to rival any sloth. Most interactions with the latter type are simply irritating, but some have brought me to the brink of losing it completely. For our part, we have of course a clear understanding that we are dealing with people who have not had any formal training, which means that certain tasks, especially any that involve handling sea turtles or their eggs directly, should not be carried out by a volunteer. There are, however, many simple tasks volunteers can learn in an afternoon and carry out independently. Unfortunately, some volunteers disregard instructions, can't be bothered to put in much effort, or completely mess things up.

What probably annoys me most are volunteers who are in party mode because they are on vacation while I'm trying to get my work done. More than once, helpers have stayed out late boozing in one of the local bars and either arrived for their shift completely plastered or didn't show up at all. I have lost count of the times I have had to find replacements at the last moment or have had to jump in myself even though my own shift has just ended. I wouldn't mind so much if their thoughtlessness wasn't sometimes accompanied by a certain indifference toward or lack of empathy for the turtles. Unfortunately, some volunteers really do come mainly to party or take selfies with the baby turtles. Beyond that, they're not much interested in the work. I reached breaking point with one volunteer I found lying in the shade at the hatchery in the midday heat. When I checked the protective cages around our nests, a number of babies had already hatched in two of them and were lying motionless in the hot sun. I immediately opened the cages, brought the turtles into the shade, and put them in our cooler with cool, damp sand. Then I ran, incandescent with rage, to the volunteer and blew up. She was supposed to check the cages at least every twenty minutes for hatching babies and take them out right away so they didn't bake in the sun. Clearly she hadn't checked for hours because she was hot and wanted to stay in the shade. Because of her, the turtles almost died. I can't grasp how anyone can have so little sense of responsibility and a complete lack of compassion. Over the years such experiences, although they are in the minority, make many of us regular workers less than enthusiastic about dealing with volunteers.

Apart from being an opportunity for tourists who are interested in meaningful work and a stepping stone for future conservationists, voluntourism is often the most important source of income for many projects. The money pays for research equipment and staff wages, and local families benefit as well, because they are usually the ones that provide food and lodging for the volunteers. That, in turn, leads to fewer problems with poaching. When this kind of tourism began to become popular in the 1990s, volunteers usually contacted the projects directly, but as more people became interested, specialized agencies soon realized there was good money to be made. They systematically insert themselves between projects and people interested in volunteering, supposedly acting as go-betweens to connect the two sides but effectively acting as gatekeepers, because they have more staff and more know-how in marketing matters and search engine optimization. Volunteers often pay high prices thinking this money is going straight to the conservation projects. But the sad reality is that the largest chunk goes into the pockets of these for-profit intermediaries while the projects continue to struggle just to survive. The sums paid to the projects by these companies usually don't cover much more than the costs of food and lodging for the volunteers. And so the projects must settle for quantity rather than quality while catering to the most outlandish demands to keep participants in a good mood. Too often this leads to a shift in the projects' priorities that unfortunately doesn't benefit either the turtles or the overworked regular employees. Many of the volunteers feel betrayed by the agencies when they realize where their money actually ends up. That, in turn, has led to increased scrutiny and justified criticism of this kind of tourism. Luckily, however, there are also positive examples of organizations doing this differently. Agencies such as SEE Turtles and Working Abroad have been supporting some projects financially for decades and providing amazing, hardworking volunteers.

No matter how one views voluntourism, when organized well, it is important for conservation today. For decades, it has been the only way for many projects to carry out their work, and it has given thousands of people from all over the world a new appreciation for nature, sea turtles, and what we do. As frustrating as the interactions can be sometimes, I really appreciate that this kind of volunteer work has created something like a global "Dumbledore's Army" for sea turtles that is always ready to support initiatives, that believes in us, and that has our back. I am still in contact with many of my former volunteers or at least follow them on social media to see how, over the years, young high school graduates and college students have matured into conscientious adults. My heart always beats faster when I see that their time spent with sea turtles in Costa Rica left its mark on most of them. Even if they do not go on to pursue a career in conservation, they still carry the values they internalized here into their private and professional lives and pass them down to their children. They are part of a tight-knit community whose presence makes me feel less alone when I tackle daily issues around sea turtle conservation. Their passion and enthusiasm, their faces and life stories, build me up when I am down as I am swept along with them. It gives me the courage and strength I need to keep going when I know that so many people believe in me and the work that I do.

I recommend that all prospective conservation biologists volunteer as early as possible to become part of this amazing community. The opportunity is also a great reality check on people's expectations before they decide once and for all to embark on this career path. It's most informative—and in some cases it's a wake-up call—to do hands-on work in a conservation or field project. Otherwise, after years of working toward their goal, a person might discover too late that this kind of work is not for them after all. I have a few colleagues, although they would probably never admit it, who find the idea of researching sea turtles more attractive than the fieldwork itself.

At this point it is good to keep in mind that scientists who do not work in the field also make extremely valuable contributions to sea turtle conservation. Whether they work in a laboratory with blood and tissue samples researching sea turtle genetics, or model population dynamics on a computer, or generate distribution maps, all that work is relevant and immensely important. I should also mention that research can be done in many different ways; it is not always carried out at universities by PhDs and professors. I was brainwashed into believing this narrative when I was a student, and I still hear it from college students. The world is a big place and there are so many opportunities both within and outside of the scientific community to work with and on behalf of sea turtles. Sometimes you have to think outside the prepackaged boxes our society offers and look elsewhere to get inspired and find your own path. But it's worth it, and our sea turtles need all the help they can get.

Professional conservationists, meaning biologists and graduates of programs in biology and other environmental sciences, are often highly critical of voluntourism because they feel exploited when they are expected to work for free. Volunteers are not usually paid and, in the case of voluntourism, volunteers actually have to pay to work, which means these professionals are not completely wrong. Voluntourism, however, is not meant for them, but for people from other walks of life with no prior knowledge of the subject who want to make meaningful use of their vacation, their time, and their money. I should also mention that this criticism comes almost exclusively from professionals in the Global North where young people, even those who are disadvantaged in their own countries, are way more privileged than people from countries with less economic development. They expect to be paid European or North American salaries for work in a tropical country, which shows that they probably haven't thought much about the realities of living there. When they arrive, they take away jobs from locals, jobs that are enormously important both for the people that live in these countries and for the objectives of sea turtle conservation. Accordingly, projects that are mindful of what they are doing always give preference to local people when they hire paid employees.

I feel I should also mention that most biologists fresh from university still have a thing or two to learn about fieldwork. Most universities don't prepare their students well enough or at all for the kind of work that we do. Therefore, unfortunately, recent graduates are often of limited use to us. Anyone who wants to be a field biologist needs to first get practical experience and competency elsewhere—for example, through unpaid volunteer work or internships. It's not fair, of course, especially for students in the U.S. who have already paid thousands of dollars for their education. If there were more money for conservation generally and this money were distributed more equally, all of this would probably not be an issue. It would also allow projects to build a stable team that could be trained and developed over the years. But as long as neither of those things is a reality, I can understand projects that use voluntourism to pay their local employees and rely on volunteers for unskilled labor.

As a project leader, I see the problem with funding my work from voluntourism more in my dependence on an extremely unpredictable source of income. The number of tourists fluctuates wildly, and because I never know exactly how many paying volunteers I am going to get, I never have the firm financial footing I need to plan. How many paid employees can I afford this season? Will I be able to pay them for the whole season? The coronavirus pandemic highlighted this issue when all of a sudden volunteers could no longer travel here. Many projects had to put their work on hold for one or even two years, and quite a few never recovered.

MY DREAM IS that one day conservation will be recognized and valued—including financially—for what it is: a service to humankind. Healthy ecosystems with a healthy biodiversity are the basis of our existence, and having both in the long term would ensure that humanity itself would be healthier. Pandemics could be avoided in the future if we stopped destroying habitats and restored them instead so wild animals had enough space to live their lives in peace. Governments and investors all over the world would do better to put their money into conservation instead of subsidizing conventional agriculture or drilling for oil. That way all of us in nature conservation could get the financial support we so desperately need and carry out our work without having to worry about money. There would be more paid positions and we could pay all professionals a fair wage, and it would be more financially attractive to follow a career in nature conservation.

But that is not how things are. And so, in my line of work you still find mostly privileged *white* people from the Global North who are either financially independent and can afford to earn less, or who have simply decided to make do with very little. Unfortunately, this fact often leads to conflicts and misunderstandings with local communities, for which many conservationists have very little compassion and understanding. There is simply no common ground because life experiences and living conditions are so different. I believe the solution is to train and encourage local experts who will be able to unite these two very different worlds. Only when the people who live in these countries are on board can conservation become truly sustainable. I hope this beautiful dream of mine will one day be realized.

To realize the dream and bridge the gap, our project pins its hopes on community outreach and environmental education. In most cases, this means weekly visits to local schools and repeated outreach events. I regularly speak to young people. My goal is to bring them closer to the mystery of sea turtles. I explain the role of turtles in marine ecosystems and communicate the impact and importance of protecting them. On more than one occasion, one of our future staff was among the children listening wide-eyed to our stories, asking inquisitive questions, and enthusiastically participating in the games and activities we had prepared.

My former mentee and colleague Ariana Oporta is such an example. She went to primary and secondary school in the tiny Caribbean village of Gandoca. Thanks to our sea turtle project, she quickly discovered her passion for turtles and turtle conservation, and was the first and so far the only person from her village to study marine biology. I got to know her when she was fifteen and have had the privilege of accompanying her on her journey for the past sixteen years. Today, I am incredibly proud of her and the competent scientist, wonderful young woman, and fearless leader she has become. My heart jumps with joy when I see the passion and empathy with which she represents our organization, COASTS, as its president.

Ariana gives me hope that there is a new generation coming who will pick up the baton as I get older. I have a bum knee, which for a few years now has made it difficult for me to walk in soft sand. And years of sleep deprivation have taken their toll. These days if I have one sleepless night I feel as though I have been awake for months and the effects can last for days. Given all that, I am fully aware that I need younger assistants and eventually someone to take over from me. I'm already thinking about that because we can only achieve the best and most sustainable outcome for the turtles if the local population is on board with continuing these conservation initiatives over the coming decades. It often takes a new generation to shake up the system. We need to be in this for the long haul, and we need to have competent people who are willing to make the necessary sacrifices.

The next generation also needs to be more diverse so we can come up with different ideas and find new, creative approaches to solving the problems we face. We urgently need to recruit people from underrepresented groups, but if we are to do that, some things in the working conditions surrounding nature and sea turtle conservation first need to change drastically. Racism, sexism, and poorly paid positions discourage many competent people. We desperately need more Indigenous and local voices to help break down neocolonial structures, and more women to stand together as a community committed to combating archaic patriarchal structures. In general, I hope there will be people who engage with conservation with their hearts and minds, who don't allow negative experiences to get them down, and who can focus on the positive.

Optimism is a choice. I would describe myself as critically optimistic, because I am not delusional and I am also a pragmatist. I firmly believe things can be better in the future—if I didn't, I might as well give up right now—but I don't live with the delusion that things will get better on their own. We still have a lot of hard work to do. But with the help of a new generation that still cares and is willing to fight for their future on this planet, we can do it. At the end of the day, it's not just about saving sea turtles or our oceans. It's about protecting the very things that make life possible for us on this planet, and therefore it's about the survival of humankind as a species. Young people are the ones who still dare to dream of a better world and change. It is easier to see clearly when your worldview has not yet been clouded by the cares and worries of being an adult.

But although the next generation appears to be the light at the end of the tunnel, we can't put all the responsibility onto their shoulders. We have arrived at a point in human history where we must all decide if we are going to be part of the problem or part of the solution, whether we will fight for a future on this planet or give up and do nothing. Personally, I don't want to move to Mars. I want to still be able to live on our blue planet. If we can develop innovative technologies that make it possible to colonize Mars, shouldn't we also be able to come up with new ideas to solve problems here on Earth?

I have had the good fortune to work with amazing people from all over the world who are most definitely part of the solution. I have had the great privilege of welcoming young people into my project and educating them, and in this way I have incidentally contributed to the creation of a global community for nature conservation that reaches into every conceivable sphere of life. This new generation of sea turtle conservationists, together with the new generation of sea turtles, gives me hope that there will still be sea turtles in the near and distant future and that conditions in our oceans and on our planet will improve. I'm convinced that we just need the will to change and to work hard enough to make these changes happen.

EPILOGUE

I'M STARTLED FROM SLEEP just after four in the morning. My student assistant Megan is at my door, dripping wet in the tropical downpour. A leatherback has nested but she has crawled back into the water without the night patrol noticing, leaving her well-camouflaged eggs behind. Now the assistants on the beach need my help—or rather the help of my dog, Fiona.

I adopted Fiona from a Costa Rican animal shelter at the beginning of the pandemic when she was a year old. She spent the first few months of her life on the street and is a beautiful mutt who, according to her DNA analysis, is a mixture of nineteen different dog breeds. If her genetic testing is to be believed, she is one-quarter German shepherd, followed by chihuahua, rottweiler, chow chow, collie, Peruvian hairless dog, and thirteen other breeds.

Adopting Fiona fulfilled another of my childhood dreams. Apart from my aquarium and my sister's budgerigar, I grew up without pets because my mother has a phobia of anything larger than a budgie (and even that frightened her when it flew around). Later I lived such an itinerant lifestyle that for years it didn't seem fair to have a dog. First the pandemic, and then the prospect of moving into my own house and being able to spend most of the year there, finally convinced me it was time to have a four-legged friend. For me, it was love at first sight when I met Fiona. She accompanies me on most of my adventures now and, of course, is also by my side at work.

One day we were walking together along the beach looking for nests that had hatched, when suddenly Fiona sniffed around and became intensely interested in a spot in dense driftwood. Hurricanes Eta and Iota had washed up massive amounts of debris and driftwood on the beach directly in front of nests that were still incubating. When I looked to see what Fiona was so interested in, I could make out tiny flippers fluttering about unable to find solid ground. A hawksbill hatchling had got lost in the driftwood and was now stuck. I was over the moon and praised Fiona effusively for her find. But that did not satisfy her. She went back to running around among the pieces of driftwood, her nose to the ground, until she once again began scratching excitedly with her paws. I ran to her—and there it was, another baby sea turtle. After I had praised her enthusiastically once more, she was up and off again. In the end, Fiona found seventeen live babies in the driftwood and five dead ones.

We returned home in a euphoric mood, and along the way I came up with a plan. I would contact Chris Fritz. I had met Chris and his dog, Saul, a few months earlier at a sea turtle conference in the U.S. Saul was a search and rescue dog that was also trained to find sea turtle nests. I hoped that Chris could help me train Fiona to sniff out lost babies and nests. Then she would have a job at my side. No sooner said than done. A few weeks later, Chris came to Costa Rica, and we began training. Fiona, like me, was an attentive student, and by the time Chris left, she had gotten the hang of the sniffing game.

Since then Fiona and I have fine-tuned our skills, and when my assistants are having difficulty finding a nest, they come to us. Fiona is the piece that had been missing from my life—a new, positive counterbalance to my work. I often joke that Fiona is my therapy dog; she always cheers me up when I am down, because sometimes I am weighed down by worries about the future of our sea turtles.

More than sixteen years have passed since I saw my first leatherback turtle nesting. Although my life constantly evolves, it still revolves mainly around sea turtles. Much has changed since that day, and even though there is some good news in matters to do with sea turtle conservation, the overall trend continues downward. From the eight hundred leatherback turtle nests on the beach in Gandoca during my very first nesting season, numbers have dwindled to now only around fifty each season. Similar declines have been observed on other Caribbean beaches. For that reason, the Northwest Atlantic leatherback turtle subpopulation, including the Caribbean nesting population, was reevaluated and reclassified as endangered on the IUCN Red List in 2019, after the global population of leatherbacks had been rated as "only" vulnerable in 2013. Sea turtle numbers across the globe are declining, and many populations are 80 to 90 percent smaller than they used to be. A few populations, such as the leatherback population in the Eastern Tropical Pacific, are literally on the brink of extinction. It pains me to think that in the not too distant future, these magnificent mother turtles may no longer be visiting my beloved Pacific beaches.

The inhabitants of our oceans are beset by threats that appear to be utterly overwhelming. In conservation biology, therefore, we often talk of the biblical horsemen of the apocalypse to illustrate the serious dangers of the increasing and combined pressures now threatening the environment. Many of these threats are complex webs of cause and effect that interact and sometimes create deadly synergies. In general the four horsemen are (1) habitat destruction; (2) overexploitation of resources; (3) pollution, including pollution that drives climate change; and (4) invasive species. These four are considered to be the main drivers of biodiversity loss and species extinction. But these simplified descriptions don't capture the complexities of the situation. To get a better understanding of how this all works, it helps to take a quick look at sea turtles.

Sea turtle nesting habitat is slowly disappearing as coastlines are developed. It is also disappearing as sea levels rise in response to climate change. Excessive—now illegal—harvesting of eggs and poaching of animals for meat and tortoiseshell over past centuries have pushed numerous populations to the brink of extinction and have prevented population recovery in many tropical countries because not enough babies are being produced to replace turtles that die. Global temperatures continue to rise, meaning that on most nesting beaches mainly females are hatching. Industrial fishing in recent decades has systematically overfished our oceans, and hundreds of thousands of sea turtles die agonizing deaths as incidental bycatch. Ghost nets that float untended in our oceans as a direct result of our fisheries kill even more indiscriminately. On top of all this, most nets are made from plastic, a material that directly contributes to the climate crisis through its manufacturing process and causes dramatic health problems and unimaginable suffering as it drifts through the ocean in the form of fishing nets or plastic bags. Invisible pollution from agricultural fertilizers causes dead zones in the ocean where there is no oxygen and nothing can live. Increased nutrient levels lead to massive blooms of algae that sink and decompose, a process that in turn depletes the supply of oxygen in the water, impacting the health of marine life. Not only that, but these fertilizers accumulate in marine plants and animals, leading to diseases such as fibropapillomatosis. In other places, invasive species, such as the seagrass *Halophila stipulacea* in the Caribbean, are replacing the turtles' natural food sources. It's safe to say that sea turtles are being bombarded from all sides.

Against this backdrop of complex and convoluted connections, the details of which are sometimes unclear even to scientists, there has been a trend in recent years to simplify the picture by isolating individual threats as the "main problem"—and ignoring the rest. People go out of their way to find a magic bullet that will not only solve the "main problem" but also all the other environmental issues in one fell swoop. It might make some people feel better to break our world down into smaller, more easily digestible chunks, but these simplifications in no way reflect what is really happening on our planet and in our oceans.

Dividing problems into small parts might make our escalating global environmental crisis appear more controllable. Our efforts become easier to keep track of, more focused, more manageable, and we feel less overwhelmed by all the problems we are facing everywhere in the world and in our society. But we will not be able to save our oceans by no longer eating fish if we continue to produce and discard plastic, use fossil fuels to produce energy, raise an overabundance of cows for milk and meat, and continue to practice factory farming that depends on monocultures, fertilizers, and pesticides. Similarly, we won't prevent sea turtle populations from becoming extinct by giving up plastic straws but otherwise continuing to live our lives as we did before.

However, we have to admit that we cannot fight in all corners at once and that we have to choose our battles wisely because all of us have only a limited amount of time and energy at our disposal. But no matter what we are passionate about, we should always respect that other people and organizations are doing work that is equally worthwhile in other areas, even when their approaches and passions are different from ours. We can solve our many problems only when we work together: the scientific world, conservation organizations, and individuals. We no longer have room for big egos focused on name recognition who monopolize data and money. It's about something bigger than ourselves. Unfortunately, we rely on conflict instead of cooperation far too often. But nobody is going to save the sea turtles, end the climate crisis, or win the battle against plastic by themself—we can do that only if we work together.

Of course, we must hold manufacturers, politicians, and individuals who have an inordinately large influence on decisions and emissions accountable, but that mostly lies out of our direct control. As individuals, we can influence the trajectory of climate change, overfishing, environmental destruction, and ocean pollution by our behavior. What we buy, how we live, and our demands as consumers influence the products we are offered and the decisions manufacturers make. Beyond that there are many concrete actions we can take to help sea turtles, either directly or indirectly.

- We can work together to combat climate change by reducing our emissions—flying and driving less, reducing or completely eliminating our consumption of dairy products and meat, and cutting down on our use of plastic. We can choose energy providers that produce energy from renewable resources and, of course, elect politicians and political parties that care about the climate crisis.

- The best way to combat the overfishing of our oceans and the resulting dangers of bycatches and ghost nets is to stop eating fish from the ocean. If we don't want to give up ocean fish completely, we should catch them ourselves or at least know that they have been caught sustainably.

- Illegal poaching of sea turtles and their eggs happens mostly in tropical coastal areas where the turtles feed and nest. We can support the small communities there with tourism and voluntourism to provide people with alternative sources of income to combat the need for poaching. And no matter how tempting it might be to try something exotic, we should never eat sea turtle eggs or meat.

- We should not purchase jewelry made of unknown materials at local markets, especially in countries where hawksbill turtles nest or feed, because we could inadvertently buy something made of tortoiseshell. A simple test is to hold the item to a flame: plastic melts while keratin chars and stinks like burned hair. If the colors on the piece are uniform, it is likely made of plastic. To check if the piece is made of cow's horn or tortoiseshell, hold it up to the light. The light patches in tortoiseshell are almost transparent; in cow's horn they are cloudier. Also, the keratin fibers in cow's horn have a linear structure, whereas those in tortoiseshell are circular.

- We can avoid polluting our oceans by no longer supporting the use of fossil fuels. Instead of buying the products of industrial agriculture, we can frequent local organic markets that sell products grown by farmers who avoid monocultures and grow their crops without synthetic fertilizers or pesticides. And we must drastically change our relationship with plastic by avoiding single-use plastics and reusing other plastic products as many times as possible so less plastic ends up in landfills. As we vote with our wallets, we should support companies and manufacturers that aspire to develop more ecofriendly packaging instead of deceiving us with greenwashing. We must find ways out of our disposable economy and toward a circular economy in which ideally 100 percent of resources and materials are recycled.
- When visiting sea turtle nesting beaches on vacation, we can take care to book only those hotels that have special lighting systems during nesting season. We should avoid countries and regions that allow unregulated development of coastlines. We should never use

white light at night on nesting beaches, and at the end of the day, we should remove all beach furniture and beach umbrellas, fill in all holes, and level all sandcastles.

- In general, it always helps to support organizations and individuals fighting for the conservation of sea turtles. Whether they are doing so by protecting females and their nests on beaches, by protecting turtles in all stages of their life out in the ocean, by securing and restoring nesting beaches and feeding grounds, or by establishing new protected areas—all these organizations and individuals need our support.

We can't wait for someone else to do the work for us. We all have to pitch in. If you take just one message from this book, I hope it is the realization that sea turtles are worthy of our protection, and that each of us can make a contribution toward their survival and make a difference every day. Even if it sometimes seems overwhelming or you feel alone and insignificant because no one else in your family or circle of friends is with you, I assure you: we are many and we are everywhere in the world!

ACKNOWLEDGMENTS

EVER SINCE I LEARNED TO READ, it has been one of my lifelong dreams to write a book, but I always thought you had to be very old and wise to write down what you knew and parts of your life journey. Luckily, my wonderful editor Ann-Marie (Mia) Mecklenburg saw things differently, and I am grateful from the bottom of my heart that she "discovered" me and believed in this book. Quite apart from that, she is also an excellent editor and steered my creative outpourings into the right channels—or you could say, along the right lines—with great skill and tact. Mia, you are the best! At this point I would also like to say a big thank-you to Christian Weigand from the podcast *Helden der Meere* (Heroes of the seas) for inviting me to talk to him, because without him Mia would probably never have heard of me.

In the beginning I told only a small group of people that I was writing a book, so my husband, Andrey, my good friend Elise Mattison, and my mutt, Fiona, were the ones who bore the brunt of the emotional impact writing this book had on me. From the bottom of my heart, thank you. I would never have managed without you. As words piled up into chapters and chapters piled up into a book, I let more people in. I would like to thank my friend Dominika

Berger, who stood by me with advice and support on the homestretch. I would also like to thank my sisters, Caroline and Ulrike, and my parents, who allowed me to bug them with questions and who shared my excitement as the project progressed. A special thank-you goes to my 2022 COASTS sea turtle team. As I was writing this book while also working for the Footprint Foundation, my colleagues in Gandoca often had to manage without me during the nesting season, which they did superbly. I am immensely proud of everything you peeps achieved!

This book would not exist without Malik Verlag, and I thank everyone there for the opportunity to tell people about my turtles and hopefully to inspire them. An extra thank-you here to Fabian Bergmann, who gave the words in this book a final polish. I thank Greystone Books in Canada and Jane Billinghurst for the English edition of my story.

When it comes to information about sea turtles, I stand, of course, on the shoulders of giants: scientists who, since the days of Dr. Archie Carr, have been tirelessly researching these fascinating creatures and sharing what they have learned. I'd like to give special thanks to Lutz Prauser, Dr. Dave Owens, and Dr. Kenneth Lohmann, who willingly shared with me their knowledge of sea turtles as food, the pineal gland and its function, and how sea turtles navigate using Earth's magnetic field.

Beyond the confines of this book, I would especially like to thank Tino Scraback, Friederike Nolting, and Sonja Renschler for their friendship and trust in me and my work. Without them and their organization, ProMar e.V., many things would not have been possible in the past few years. I would also like to thank Tropica Verde e.V., Milkywire, and Hannes Jaenicke's Pelorus Jack Foundation for their support of our work in Costa Rica.

Finally, I would like to thank all those who always stood by me over the years, encouraged me, or simply listened. Especially my fellow "Münsterites" Wolfgang Brunner, Marion Schenner, and Jörg Feldhoff, because without you my teenage years would have turned out very differently; my professor, Dr. Nico Michiels, who made studying biology fun again; my diploma thesis supervisor Dr. Heike Feldhaar, who gave me the opportunity to return to Costa Rica and undertake my first independent research project on sea turtles; Dr. Jochen Drescher and Dr. Ina Heidinger, without whom I would never have finished my diploma thesis; George Halekas, who offered me a sympathetic ear and wise counsel at times when I didn't know what I was going to do with my life; Rena Gneo, who pulled me out of a few dark places with her positive attitude and unwavering support; my two emotional rocks in Costa Rica, Magali Marion and Ana Eugenia Herrera; Susan Koehler, la familia Ramirez, Dr. Richard Woodward, and Rosibel Woodward, who have been pragmatic optimists and reliable cheerleaders over the past few years; my College Station crew Dr. Lauren Redmore, Dr. Jordan Rogan, Diana Solis, Dr. Brie Myre, Dr. Ishita Chandel, Dr. Lee Fitzgerald, Dr. Duncan MacKenzie, and Jennifer Bradford, who made life in small-town Texas more than bearable; and last but not least, my doctoral advisers Dr. Pamela Plotkin and Dr. Joseph Bernardo, who over the past ten years have had a profound impact on my life and have always had my back. Thank you!

INDEX

Note: *page numbers in italics refer to illustrations.*